Weight-Loss Apocalypse

Emotional Eating Rehab
Through the HCG Protocol

Robin Phipps Woodall

authorHOUSE®

AuthorHouse™
1663 Liberty Drive
Bloomington, IN 47403
www.authorhouse.com
Phone: 1-800-839-8640

First published by AuthorHouse 11/22/2011

ISBN: 978-1-4678-4563-2 (sc)
ISBN: 978-1-4678-4564-9 (e)
ISBN: 978-1-4678-4565-6 (hc)

Library of Congress Control Number: 2011919732

Printed in the United States of America

*Any people depicted in stock imagery provided by Thinkstock are models,
and such images are being used for illustrative purposes only.
Certain stock imagery © Thinkstock.*

This book is printed on acid-free paper.

Dedication

To Dr. ATW Simeons for his curious mind, his lifetime of scrupulous observation, and his steadfast belief that the body isn't as naïve as we think.

To the scientists who've dedicated thousands of hours exploring the curiosities and intricacies of the human body—not because they want personal gain, but because they are seeking what is important to humanity. It is their work I used to create a hypothesis, not mine.

To all of the people who've personally shared their hCG protocol experience with me on many levels. They endured poking, pinching, tests, and assessments, and submitted themselves to my brutal honesty on a weekly basis, for months at a time.

To the most important people in my life: My husband Mark and my children Chloe, Wyatt, and Suzanne. *You brighten and magnify the light of my soul with your unconditional love.* Thank you for believing in me, and for giving up everything to move across the country to prove it.

Contents

SECTION 3

The HCG Protocol—What Every Doctor, Participant, and Skeptic Needs to Know

Foreword

By Mayer Eisenstein, MD, JD, MPH

It is rare that you come across a scientific theory that is both simple—and profound. Usually the theory is so esoteric that it is not understandable, or so simple that it is not applicable. In her book, Robin Woodall has accomplished both these objectives, explaining her theory 1) about what causes obesity, and 2) its possible cure.

Woodall is an Exercise Physiologist, Certified Strength and Conditioning Specialist, and is certified through the American College of Sports Medicine. Bringing together all she learned from those specialties, she has worked in the exercise and weight loss industry over 13 years.

In her book, Woodall explains with simple but brilliant words, how the obesity epidemic is caused by eating "without hunger." It seems simple enough, but that alone will not solve the problem. She also explains how using the modified hCG protocol, as outlined by the late Dr. Simeons in his manuscript *Pounds and Inches*, can go a long way toward solving the problem of obesity and Metabolic Syndrome. Metabolic Syndrome consists of some or all of the following: elevated blood pressure, elevated cholesterol, elevated triglycerides, insulin resistance, and central obesity.

Following a protocol by Dr. Simeons, which consisted of low-dose Human Chorionic Gonadotropin (HCG), along with a very low-calorie diet (VLCD), his patients had enormous success for the treatment of obesity and Metabolic Syndrome. In his documented findings, he purported that his HCG protocol would cause weight loss, and reduced inches, by mobilizing abnormal fat stores in the abdomen, neck, arms, and legs. By doing so, it would lower or eliminate the need for pharmaceuticals to treat the symptoms of Metabolic Syndrome.

Today the statistics for weight loss are grim. Less than 2% of people who are 50 pounds overweight will lose the weight, and even grimmer, less than 2% of the ones who lose the weight will maintain the loss for at least one year. That means less than 1 in 2,500 will maintain a 50 pound weight loss for one year. Even grimmer, if you are more than 100 pounds overweight, the chances of maintaining a 100-pound weight loss for one year is 1 in 10,000. My undergraduate degree is in science, with a strong emphasis in statistics. I know from my educational background you cannot beat the odds. But, after reading Dr. Simeons' manuscript I was intrigued to learn more.

After a great deal of study, in September 2009, at the age of 63, 6' tall and weighing 334 pounds, I chose to do a slightly modified version of Dr. Simeons' hCG program, using prescription sublingual HCG that a compounding pharmacist prepared. The modification consisted of adding nutraceuticals* (Vitamin D, probiotics, multivitamins with minerals, and digestive enzymes). All of these nutraceuticals have been shown to aid in weight loss in a calorie-restricted program.

By Christmas 2009, I was down to 275 pounds, my lowest weight in more than 30 years. For the first time in over 20 years, I had normal blood pressure readings, and no longer needed prescribed medications to control it. And—the weight on my driver's license was *more* than my actual weight.

With my personal success and affirmation of the value of the protocol, I integrated the hCG program into my practice in March 2010, using Simeons' Protocol with the nutraceutical modification. At the Metabolic

---

* Nutraceutical is a term that comes from the combination of the words nutrition and pharmaceutical. The word is meant to describe natural food-based (nutrition) substances that have pharmacological (healing) effects on the human body. Nutraceuticals refers to any food product, supplement or dietary substance that has shown to have health and medical benefits.

Syndrome and HCG Weight Loss Clinic, I had a dual purpose: I would be the inspiration for my patients, and they would be mine. Today I've lost over 100 pounds using the hCG protocol, reducing my body weight to 225 pounds.

To date we've had approximately 1,000 patients use this protocol, with more than 800 staying with the program—collectively, losing over 20 TONS! Like me, many of these patients have *reduced* their need for prescription drugs for controlling blood pressure, cholesterol and blood sugar, and others have *eliminated* the need completely, as their metabolic values have become normal.

My 20th patient was just added to the elite club of those who lost over 100 pounds—an incredible achievement. More than 300 patients have lost more than 50 pounds, and many who have lost between 20 and 50 pounds. Like me, many have also maintained the weight loss for over a year.

But until recently, I hadn't been able to lose the last 30 pounds. I consulted with physicians, pharmacists, exercise physiologists, personal trainers and dieticians, all to no avail. I had become complacent at my new weight until I called Ms. Woodall for a consultation. Her answer was both simple and brilliant: "Dr. Einstein, are you eating only when you are hungry"? I told her no; I would eat even if I was not hungry. She then said, "You are still eating too much." When Ms. Woodall explained her theory, I was intrigued, but not totally sold. Upon her advice, I began to eat only when hungry. It seems simple, but this, along with the hCG protocol, is very effective. I am pleased to report that I have started to lose weight again.

In *Weight Loss Apocalypse*, Ms. Woodall has put the final piece to the puzzle, and explained why the hCG protocol, along with a very low-calorie diet, works. She also reintroduced a simple but profound concept: eat only when you are hungry.

Dr. Mayer Eisenstein, MD, JD, MPH, is a graduate of the University of Illinois Medical School, the Medical College of Wisconsin School of Public Health, and the John Marshall Law School. In his 39 years in medicine, he and his practice have cared for over 75,000, children, parents, and grandparents. He is the founder and Medical Director of the Homefirst® Health Services. He is Board Certified by the American Board of Public Health and Preventive Medicine, and the American Board of Quality Assurance and Utilization Review Physicians. He is also a member of the Illinois Bar.

His latest book, *Making An Informed Vaccine Decision* goes along with his other books: *Give Birth at Home With The Home Birth Advantage*; *Safer Medicine, Don't Vaccinate Before You Educate, 2nd Edition*; *Unavoidably Dangerous-Medical Hazards of HRT*; and *Unlocking Nature's Pharmacy.* Some of his many guest appearances include: "The Oprah Winfrey Show" and "Hannity and Colmes."

Introduction

I first heard about the hCG protocol from my sister, a registered dietitian. She'd already finished her first week on it before she called to tell me about her experience. I'm sure she waited because she expected me to criticize what she was doing.

You could say I was skeptical, considering I'm a personal trainer, and my college degree is in exercise physiology. Obviously eating less than 500 calories in food causes weight loss, but what about muscle loss, too? Destroying your metabolism? And preparing you for what? You can't eat 500 calories the rest of your life! Yes, I was critical.

At the time, I thought I was thoroughly informed as to how the body uses fuel, and how the body responds to starvation. Before I read Dr. Simeons' 1967 manuscript, *Pounds & Inches*, I assumed it was written by a con man taking advantage of our desperation due to the obesity crises. But to my surprise, the manuscript made some sense. I could relate to his theories because his observations of fat gain and loss paralleled my own during three pregnancies, and what I'd witnessed with many clients over my ten-year career.

I'd observed, measured, and assessed body fat compositions for thousands of people. Some of those clients meticulously exercised and reduced their food intake—without results. I watched as female clients gained abdominal fat during menopause, even though they were eating less and exercising more. I observed clients before, during, and after pregnancy, and witnessed the shape of their bodies change, adding fat in some areas, and losing it in others. I knew through experience that fat gain and loss were linked to hormones.

I was excited when Simeons' observations validated mine.

Dr. Simeons assessed scale weight for decades to deduce the strict guidelines of his protocol. He had theorized, based on his observations, that the pregnancy hormone, human Chorionic Gonadotropin (hCG), somehow prevented symptoms of starvation during a 500-calorie protocol. In 1967, he privately published his findings and hypothesis in a manuscript called *Pounds & Inches: A New Approach to Obesity*. Dr. Simeons' believed the brain determines where and when fat is used for fuel. He observed that hCG redirected the brain to use abnormal fat (fat that's difficult to lose) for energy. He observed that 1) participants felt minimal hunger, 2) their weight loss was rapid, and 3) their losses were specific to areas that regular diet and exercise didn't influence.

Dr. Simeons was convinced that by tricking the brain with hCG, and manipulating fuel demand with the 500-calorie protocol, the brain would "re-set" its fat-burning capacity. Ultimately, this would allow the participant to eat normally without having the same susceptibility for fat gain when the protocol was over.

After understanding his protocol, and finding a doctor willing to prescribe it, I decided to present the information, as well as the opportunity to do the protocol, to appropriate clients. They had to allow me to follow, measure, and record their progress, and in particular, compare their metabolic rate before they started, and after they finished.

Six clients agreed to participate. Before starting the protocol, I did a battery of tests. These included a record of: metabolic testing to find out how many calories they burned in a day, a cardiovascular endurance test done on a treadmill, blood pressure, resting heart rate, flexibility testing, push-up and sit-up tests for muscular endurance, bench press to measure their estimated strength, two different body-fat composition assessments, as well as circumference measurements.

I continued to measure everything weekly during and after the 500-calorie protocol, except for the fitness testing and metabolic rates, which I measured again at the end. I wasn't surprised by their significant fat and size loss but I was not prepared for the drastic amount of fat lost in the stomach area. Quickly I added three additional circumference measurements to that stomach area to ensure I was accurately assessing their size change.

All participants agreed they had minimal hunger, and most said their energy level was good. Completely shocking were the fitness and metabolic testing results after the protocol was completely over. Not only did the fitness tests improve, but the amount of calories their body burned in a day significantly increased. Considering each participant ate less than 500 calories for over a month, these results were astonishing.

I knew there had to be a logical explanation. I worked with new clients on the protocol, and continued to perform all of the tests before, during, and after the protocol. After collecting data for over 40 people, a local university statistically analyzed my metabolic testing results. When the results came back they were the same as my own observations. The metabolic rates significantly increased after the protocol. At this point, I was determined to find legitimate reasons for the increase.

After all, my findings completely conflicted with everything I was taught regarding calories, fat loss, muscles, metabolism, and what's supposed to weaken during chronic starvation.

I compiled the data, and searched for local doctors of endocrinology that specialized in organs and hormones of the body, hoping they would be interested in my findings. I assumed they'd be able to explain how the hCG could influence the body in a way that prevented typical symptoms of starvation during such drastic caloric restrictions. I wanted to know why all my participants had overall improvements, after the fact. Only one doctor was willing to meet with me. After 45 minutes of discussing what I'd been doing, and my hope for some answers, he suggested

I create a hypothesis that might make sense of what I observed. I was taken aback that he didn't have any answers, and that I would have to do my own research to understand what I'd been witnessing.

If this endocrinologist, who had a Ph.D., and also owned his own diabetes center, couldn't help, then I was definitely on my own. That evening I started from square one, searching for scientific explanations for the physiology of hunger, energy, fueling, and how the body regulates fat metabolism, starvation, etc. This was a huge undertaking that required the ability to read scientific journals and reports to understand the cellular physiology.

Fortunately, my degree focusing on physiology came in handy. However, I was not prepared for what I found—thousands of medical journals written since I graduated in 2000— describing new hormones, new mechanisms, and new explanations for how the body regulates metabolism, hunger, and its complex fueling systems.

For two weeks I spent 12 hours a day, cross-referencing, reading and re-reading material over and over. I created charts and my own dictionary of organs, hormones, and functions, basically teaching myself the new physiology of energy homeostasis/equilibrium. The difficulty was not in finding the answers, but in understanding the new terminology and mechanisms I was not taught in school, or with any of my certifications.

Once I understood the most influential hormones involved in hunger and metabolism during both starvation and feeding, I formed a hypothesis that made the most obvious sense to me. Based on the modern science I studied, hCG must stimulate sufficient leptin in order to prevent all symptoms of starvation.

I felt like I was about to answer a million-dollar question, because I was so confident that the link between hCG and leptin had to be the answer. Within seconds of entering those two key hormones into a search engine, all of my hard work and focus became worth the effort. Immediately I found studies that connected hCG to leptin. Some specifically indicated the most powerful relationship between the two hormones occurred at almost the exact amount Dr. Simeons prescribed for the hCG protocol. My heart was racing, and I literally jumped up and down with excitement.

Not only did scientific evidence exist that could easily explain how hCG, through the stimulus of leptin, could prevent symptoms of starvation during a very low-calorie diet, but a huge body of science also explained the reason for significant metabolic increases. Any scientist involved in the new studies of the hormone leptin, as it concerns starvation, fat gain, and fat loss, would find the answer obvious. The only reason a person would have minimal hunger, increased energy, and wouldn't experience lean tissue loss during a 500-calorie protocol, is if hCG adequately stimulated blood leptin levels.

I wrote a hypothesis, but unfortunately the doctor who suggested I find the answers never responded. I continued to collect data and be amazed at the protocol's physical results. Today, I've closely monitored over 500 people through protocols, and continue to follow Simeons' method. However, I use only modern science to explain how it works.

I've met with doctor after doctor, but so far *all* of them have been unaware of the new science, and most haven't a clue as to the function of leptin. The protocol "experts" on TV, or who've written books on "new and improved" protocols, have yet to even mention new science, and continue to reference Dr. Simeons' outdated and insufficient theories. It's not surprising, considering the amount of time, effort, and extensive research it took for me to educate myself to formulate an answer.

Each person I met with to discuss the protocol, continued to reference what he found on the Internet, which is based on misinformation that inaccurately explains the protocol.

But, with the lack of relevant scientific explanation, and the majority of hCG sold on the Internet by people who aren't necessarily educated in weight loss or human physiology, it makes sense to repeat the theories found in Simeons' manuscript. No other explanation is out there. Unfortunately, there are many people attempting the protocol who don't follow it the way Dr. Simeons intended.

Most participants are completely unaware that eating and hCG directly influence the hormonal response from all of the organs in the body, and without strict compliance to the protocol, there could be harmful consequences. It's approached like a diet, and businesses have started to manipulate and change the protocol to increase their profits and to make it more appealing to the masses. It's unfortunate, considering the decades of observation and work it took Dr. Simeons to deduce the specificity of protocol. But Dr. Simeons didn't really know how the protocol worked either, and even he admitted laboratory explanation and proof was needed.

If he had known what was going to happen with our culture, as it concerns over-eating and obesity, and the lack of integrity in the hCG diet industry, he would have written and presented the protocol differently Dr. Simeons didn't foresee his protocol would be used, prostituted, and misdirected as a short-term fix by the consumer. He had no idea we were going to have a massive, cultural, emotional eating disorder, and that we'd continue to blame the consequences of fat gain on everything but ourselves. Just because you have minimal hunger, doesn't mean you won't eat. Just because you have significant fat loss, doesn't mean you will be motivated to change the way you eat forever.

My goal in writing this book is to start a new conversation about Dr. Simeons' protocol that has relevance, not only as a hormonal therapy, but as a means to end our national eating disorder. Instead of continuing to apply the protocol as a short-term diet, I'd rather it be discussed as a real solution, a tool to end irrational eating for emotional fulfillment.

We are dealing with a crisis that is an addiction to emotional eating, and the obvious result is the overwhelming increase in obesity.

Think about the number of people in our culture who eat without hunger. How many people eat to gratify emotions? How many eat because they're bored? When you observe our nation's behavior with food, it's very clear that fat isn't what we should be obsessed about, and weight shouldn't be the target of the problem. We need a genuine desire to eat less, one that isn't dependent on weight loss as a reward. This requires each of us to be accountable for our own emotions, and find happiness in life not centrally stimulated by food.

- Can you find a different hobby when you're bored, instead of eating?

- Can you deal with stress without using food as a pacifier or distraction?

- Can you create happiness without having to eat?

If you can, then eating less would not be such a big deal, and you wouldn't have to pay for a diet to help. For most people, eating is almost entirely an emotional decision and behavior. If we forced the majority of our society to eat functionally, it would be torture for them, especially if you didn't allow them to monitor their weight.

Even if someone has little hunger, and no symptoms of starvation, his or her emotional distress is far worse than the physical when it comes to eating less on the protocol.

We've created a society that is so emotionally connected to eating that any form of restriction feels like punishment. Shouldn't there be some personal accountability? Look at the overall implications that emotional eating has on medical costs, health care, and the occurrence of disease directly linked to obesity. However, the decision to find emotional strength without eating must come from an individual's internal desire.

It can't be forced; otherwise we'll end up with even more psychological problems.

As more and more people choose to free themselves from emotional eating, we would see a social movement with a new cultural distaste for excessive eating. Eating minimally would be sanctioned and encouraged by peer pressure, and emotional accountability and strength would be the next "big" thing.

I believe the protocol provides the ideal atmosphere for participants to rethink the role that eating and food plays in their lives, and to develop emotional strength without needing a crutch. By letting go of food, and eating a minimum amount during the protocol, your emotional strength can be tested, and you can experience a life of eating less. However, for there to be a drastic impact on a participant's overall relationship with food, he or she has to approach the protocol with that as the ultimate goal, eliminating the need to monitor weight for motivation.

Many people falsely assume that when they lose the weight, they'll change their relationship with food.

This makes absolutely no sense, considering the role food plays has very little to do with body fat, and everything to do with their lack of emotional security. How many people have lost a substantial amount of fat, just to gain all of the weight back when they return to their "normal" relationship with food? Most.

It's time to set a new standard for ourselves, fulfilling ourselves emotionally without needing to eat for emotional support. We have to want to limit ourselves, eating less even though we don't have to. Then losing weight wouldn't matter because you'd want to continue to eat less even without a weight problem. This would require a new approach to reducing our food consumption, an intrinsic desire that no diet could enforce or create. Eating less must be a personal decision— a life-lasting change.

One of my favorite quotes is by Albert Einstein.

"No problem can be solved from the same consciousness that created it."

In the case of our culture, solving the cause of obesity would require us to allow ourselves to feel vulnerable when emotionally tested; the ultimate goal being an awakening to our own emotional strength without needing to eat. This confidence would remove the need to eat as a pacifier, and eventually the physical result would be fat loss or prevented fat gain.

The protocol provides the perfect environment to rehabilitate our emotional strength. If the hCG protocol provides a hormonal environment that indeed prevents starvation while drastically reducing the need for food, then each participant has the opportunity to re-establish his or her emotional well-being independent of food.

With the right frame of mind, participants can observe their desires to eat often have nothing to do with hunger. They can redevelop a sense of true physical hunger that would help them control their functional need for food.

The hCG protocol would revolutionize our culture. Not only because of the mass reduction in obesity, but because when people are able to develop emotional strength and well-being without needing to eat, society would be healthier, happier and more productive.

Eating less—not because we have to, but because we want to.

SECTION 1

Ending the Battle with Weight

Chapter 1

Our Nation's Battle With Gluttony

When clients mention how they "battle weight," I immediately picture a Civil War battleground. I see a large piece of land separating their obtrusive fat on one side, and their defensive efforts to restrict food on the other. Clients usually describe this battle as if the fat is attacking them. In defense, they diet, feeling their fat is an abnormal growth that just won't go away, and is plaguing them, inflicting disease, and damaging their self-esteem. It takes over how they dress, dictates their activities, and greatly affects others' negative perceptions of them, often causing people to misjudge their character.

When I ask clients what they've attempted in this battle—after they take a deep breath, it usually leads to a predictable history, starting from childhood. They've almost always subjected themselves to starving, soup diets, cleansings, liquid meals, counting calories, and/or taking stimulants and pills. You name it––they've even attempted lifestyle changes. They've participated in torturous exercise ending in injury, used fat-shaking machines, tried rubbing, injecting, dissolving, and having fat-removing or stomach-stapling surgery. Many spend extra money for packaged foods and supplements, and accept additional exercise as a strategy to attack the fat on more than one front. The fat just won't go away, or won't stop coming back. Most people consider their battle with weight a frustrating, expensive, and an obtrusive interference with their happiness, especially since dieting requires a temporary break from our cultural norm of eating without limit.

Diets can feel punishing and depriving when compared to events or activities where eating excessively is acceptable and expected as a normal part of our cultural expression of celebration—like weddings, birthdays, holidays, all social gatherings, dinner parties, vacation, family functions, sporting events, brunches, outings, or lunch meetings. However, if you're not physically famished or experiencing physical hunger, are you really being deprived or punished?

Even when food consumption is decreased, hunger is not usually why people fall off their short-term diet. As soon as motivation to eat without limitation surpasses the desire to lose fat, a diet has no value, and weight is no longer important. *Those who lost fat will eat to celebrate, and those who didn't lose weight will also eat.*

This is similar to an alcoholic who drinks when things go bad, and also when things go well. It's no wonder more than 95 percent of those who lose weight, gain it back. Diets don't resolve the emotional reasons you eat, and weight is too superficial to be responsible for your lifelong motivation to eat less.

We sabotage our battle against fat when we need a distraction, are drunk, get the "munchies," suffer break-ups or job loss, have arguments, or encounter rude people or bad drivers. We eat when we experience disappointment, lack love—or have to deal with people we love—or those we just can't stand. Even when we have nothing to celebrate or to be upset about, we eat—to relieve boredom, for entertainment, or to just pass the time. We even eat to help create any emotion we feel we need like excitement, love, good fortune, or security. It's no wonder we feel deprived by diets, and continually sabotage the battle against fat.

Having to diet requires eating less, which means having to be around others who continue to eat excessively. This is why the rate at which fat is lost is so important, indicating whether the diet is worth it. If weight is lost too slow—if it doesn't happen quickly—food deprivation can feel like punishment. And for many who actually lose significant fat, the effort it takes to continue to eat less becomes too great to maintain their losses.

Ask yourself: Would you eat less if it were not to lose weight?

We love emotional eating so much that we ignore our physical distress. We are patient with dysfunctional eating, as our clothes get tighter, we are tolerant as fat grows back, and we ignore the scale as weight increases. Why are we continually surprised by the growing trend towards obesity now affecting our children? It's obvious our battle against weight is completely missing the mark.

The question is: Why do we continue to pay the weight-lost industry to solve problems we bring on ourselves?

Temporary diets are like putting a small band-aid on a gaping wound. When they don't work, we continue to wonder why we're still bleeding. All the while, we blame our bodies for not being resilient enough to withstand the injury we self-inflict. If the way we live normally causes the problem, we need to take a hard look at why we're in denial. Why don't we blame our cultural norm? Why do we continue to waste our money, and misuse our motivation and effort with diets, when we know full well that they are only a short break away from our gluttonous norm? We are the epitome of gluttony.

By definition, gluttony is the act or practice of eating to excess.

Did you ever consider that the battle isn't against weight, but is a battle for freedom to eat without limit? In actuality, we are the offenders— emotionally justifying attacks against our bodies. Fat is the body's only defense. We want to continue our indulgence with food, and we fight to keep it. Otherwise eating less would be normal, and gluttonous desire for food would be minimal or non-existent. Our approach to dieting, and our blaming fat, is completely misdirected.

The problem is that our culture justifies emotional eating, but then discriminates against obesity. Why? We assume that the way a person looks influences and defines his or her character. We use the way we look as a visible sign of stability, sexual appeal, discipline, and accomplishment—to develop confidence, and identify our standing in society.

If you gain fat, you've crossed the line. You're then judged, and made to feel:

1) As if you have no self-control,

2) Your body has malfunctioned,

3) You are no longer appealing, when compared to our cultural model of beauty.

Our culture looks at fat as if it's an abnormal growth, and many falsely assume the more fat a person has, the more flawed his or her character must be. Instead of adjusting our self-indulgence and the fundamental reasons why we love to eat so much, we pay all our attention to how to get rid of the fat we dislike.

This demand has created a multibillion-dollar diet and weight-loss industry that takes advantage of our bias against body fat. That industry develops temporary constraints, and markets the idea of easy fat removal. We buy into it, knowing full well that when the diet is over, we're going to return to our cultural norm.

Diets assure that "heaven" will be attained if you lose the excess fat. There's over-emphasis on how much confidence you'll achieve, and how your life will change so drastically after you lose the fat that you'll be motivated forever. Thus you pay someone to ensure that you eat less.

They instill harsh judgment towards food, and you desperately submit, but only as long as you attain weight-loss heaven. You obsess over the scale every day, to motivate you to comply. But as soon as weight doesn't drop as quickly as you think it should, the food religion seems judgmental, depriving, unrealistic, and shallow. Eventually, once you've tried every known diet, you believe that all food is evil, and you're doomed to "fat hell." You might as well succumb to the fact that you're not one of the "chosen ones" who gets to eat anything she wants, and never gains an ounce. It just doesn't seem fair. Without the freedom to eat, dieting feels like hell, relative to everyone else who eats freely, and achieving a normal weight is no longer worth the effort.

Dieting in our culture is extremely difficult, and for most, it is a losing battle. When friends, family, coworkers, and everyone around you regularly indulge emotionally with food, it's hard to eat functionally without seeming abnormal. Parties suck, celebrations seem punishing, and you feel guilty when you choose to even taste the food that diets criticize and judge as "bad." Others, who seem to ridicule their efforts to eat less than the norm, can harass a dieter. There's even peer pressure to let go of a diet's restraints by others who like group feasting.

Eating is an integral part of our emotional behavioral norm. No wonder we feel deprived by diets, and continually sabotage the battle against fat.

Compared to the effort it takes to eat less among gluttons, we'd rather have immediate emotional gratification by eating with them—even if it results in physical harm. Only after fat accumulates, and our obesity becomes a more obvious result, do we again temporarily commit to another diet's restrictions. Yet, despite the fact we spend billions every year to lose weight; the trend in obesity continues to grow.

The growth in obesity isn't an indication that diets are failing, but rather suggests the way that *we eat normally* is the cause of the problem.

The problem: Our culture loves to eat without limitation more than it detests the fat. We accept any justification to eat, even though disease and death associated with obesity occur at an overwhelmingly high rate. You'd think by the quantity of food we eat that we're compensating for some sort of famine—and we are, but it's not a physical famine.

Because our emotional confidence is attached to food, we've weakened our sense of independent emotional strength. This is similar to believing lucky socks will improve your chances of winning a game. We superstitiously believe eating will improve our emotional security. As a result, when challenged by the desire to eat for emotional reasons, it's easy to believe you're not capable of creating those emotions alone.

Look at how many people stop following a diet, not because they're physically deprived, but because they feel emotionally insecure in certain situations if they are not eating. Could you have fun at a party without eating and drinking as the main source of entertainment? Have you ever wondered what you'd do if no food was available to help you create feelings of security and deal with hardship?

We're obviously emotionally addicted to eating, and are dealing with a bigger problem than an obesity epidemic. This is a nationwide eating disorder, a situation no diet can handle, especially with the misguided motivation that focuses on weight.

Until over-eating or emotional eating is exposed as the problem, our demand for excessive amounts of food will continue and weight will always be viewed as the problem. Instead of continuing to blame your body for being flawed, the food industry for inflicting gluttony, or the diet industry for failing miserably, as a culture we need to blame our rationale for unlimited eating. In the United States where food is abundant, we need to find reason to eat less, unlike in some countries where eating less is not a choice.

Can you create the desire within yourself to eat less, even though everyone else is eating more? No matter how much food you are served, can you choose to eat less because you don't need it and you don't want more? Can you live in our culture of normalized gluttony, and know that almost everyone eats too much, *and by eating less, you appear abnormal?*

You would be in the minority—a person who doesn't have hormonal problems associated to overeating and resulting obesity. You'd be saving thousands and thousands of dollars wasted on too much food, diets, and medical costs caused by disease from a self-inflicted lifestyle.

Food shouldn't be what strengthens our emotions, and neither should weight reactively motivate the way we eat. Instead of relying on a diet to work, or the food industry to properly spoon feed us, it's time we hold ourselves accountable as individuals.

- If you were given a chance to start your life over, would you eat the same way, knowing how it has affected your body?

- Would you choose to continue to rely on food for emotional fulfillment?

Find a reason to eat less that doesn't have to do with weight. Observe the behavior of others who seem to have freedom from emotional eating. People who've separated emotionally from food don't experience guilt when they eat. Eating doesn't control their feelings of security. They only eat when they're hungry, and when hunger is gone they stop eating— even if there's food left. Can you visualize what it would feel like to be emotionally free from food and dieting?

If you are unhappy with the way you eat, it's time to start over, clean your slate. Unplug and reboot.

Chapter 2

Seek, Not Hide, Emotional Weakness

Hunger is one of our body's most important senses. Unlike the sense of sight, hearing, or taste, hunger is a feeling we need in order to live. It's the one physical sensation that tells us when our body needs fuel, and as hunger goes away, that fuel is no longer needed. Without hunger's physical discomfort, we'd have no desire to eat, and eventually we'd all die of starvation. On the other hand, if the feeling of hunger never subsided, we'd continually eat without relief from its physical irritation, which would cause death.

For thousands of years, hunger was a physical cue that determined when eating was an imminent priority. As food was eaten, the cue to stop eating occurred when the feeling of hunger disappeared. This helped keep food consumption minimal, efficiently preserving food rations for longer periods of time. No one ate when bored, no one ate at specific meal times, and people were not gluttonous when they celebrated or mourned. People developed emotional confidence on their own, because excessive eating or eating without hunger wasn't an option.

People acknowledged their own emotional challenges, and developed a sense of confidence because there wasn't an over-abundance of food to use and abuse for distraction. Today, we can use and abuse anything we want, as long as we have the money to pay for it and people who tolerate it. In doing so, we've weakened our innate sense of self-esteem and tolerance for life's normal stress, and we've instead strengthened a false sense of confidence that is reliant on the person, object, or action we've used as a crutch. In our culture, *that crutch is eating.*

Today, we have such an abundance of food that we can buy it whenever, for whatever reason, and in any quantity, we want. We no longer need to farm for ourselves, preserve seasonally, or ration it among family members. Result: We've lost any reason to eat minimally.

Hunger is no longer used as important physical feedback that dictates when eating is appropriate—and when it's not.

Hunger is one of our most underrated, underused, and under-discussed physical signals. And to make it more amazing, it's built right into the body's own survival mechanisms. Using hunger as your guide is like being a child again—no more calorie counting, no worrying or shameful guilt about "bad" food, and no more self-inflicted dysfunctional gluttony.

Children don't want to be bothered to eat until they feel the physical irritability of hunger. And as soon as that hunger is gone, they're off doing what they do best, playing, creating, and having fun. Doesn't that sound liberating, to be free from the torment of dieting guilt, and the physical discomfort of gluttony?

To do this, you have to create new hobbies and find new ways to entertain yourself that don't require eating. If eating were not an option, what would you do to fill your time? Would you start or finish a project? For most people, the creative process is difficult because it requires that you let go of old ways of thinking to create something new. Holding onto old patterns of thought that sensationalize emotional reward from eating make the process very difficult. That is why many people sabotage their own efforts to change, even though the result is disease and physical harm associated to obesity.

I'm always amazed that people who've developed an addiction to eating, exercise, and dieting don't recognize how their addiction causes sadness, discontent, and unhappiness. They've developed such an emotional dependence on their addiction that they've lost touch with what it feels like to be emotionally free; to be able to create emotion without attaching those feelings to their addiction. They've relied heavily on their addiction to give them emotional security, so that without it, they feel weak, fragile, and emotionally vulnerable. This makes it easy to sabotage any effort for independence, even if obesity is the obvious result.

Case Study

People who addictively exercise to mask their emotional desires to eat without limitation usually struggle on the protocol. They suffer the loss of ego defined by the physical accomplishment that comes from the control and success of overcoming physical challenge, as well as the lack of entitled gluttony as a reward for attaining their physical goal.

One of my protocol participants was a woman tri-athlete, who was confused as to why she struggled to lose thirty extra pounds even though she trained three, sometimes five hours a day, cycling, running, and swimming. She suffered from all sorts of over-training symptoms, such as stress fractures, arthritis, chronic lower back pain, and depression. She decided to participate in the protocol because she'd promised her doctor she'd take time from training to allow her overuse injuries to heal. The protocol made sense to her, since it would allow her to reduce her weight so that her training wouldn't cause as much damage to her body.

To her surprise, losing weight alone was not enough motivation. She cheated over and over, which forced her to rethink why she was doing the protocol. With each small cheat, she'd gain weight, which conflicted with her original goals.

As she was forced to recognize the emotional distress she felt without food to mask it, her insecurities became obvious. She'd cheat for one reason, and then a few nights later, for a different reason. Before starting the protocol, she thought living with minimal hunger would be easy, but during the process she was surprised at the emotional difficulty she experienced, especially when she was socially engaged with her group of tri-athlete friends.

Now, as an observer, she witnessed the dysfunctional gluttony that all the athletes participated in, especially after they trained. Whether they were hungry or not, they ate huge amounts of food, and took delight in the fact that they didn't have to worry about their caloric intake.

Their common justification: they'd earned the right to eat because they expended the energy. Whether their body needed it or not, they were going to eat. It provided social stimulation, reward for physical attainment, and freedom to enjoy gluttony without guilt.

She came to realize that her desire to overeat had always been there. She began her triathlon career because a boyfriend broke up with her, saying he wasn't attracted to her because of the extra weight. So she exercised to lose fat, and had some initial success.

The positive reinforcement and confidence she attained through her physical achievement became an addiction. She continued to train for years. All the while retaining extra body fat, and gaining even more whenever she stopped training. Without changing her emotional need to over-consume, she ended up creating a dependency on exercise to restrict her weight, eventually not only harming herself hormonally by eating without hunger, but over-using her body, degrading and inflaming muscles, bones, and joints. Without the exercise, she had to acknowledge her desire to eat excessively. Instead she blamed lack of exercise on her weight gain.

The protocol forced her to realize that her behavior with food, and with exercise, was destroying her body, and she was justifying it so that she could feel emotionally secure. To allow her body time to heal, she had to develop emotional strength that no longer was defined by physical competition and attainment. If not, she'd feel lazy, unimportant, insignificant, weak, and unaccomplished. And worse, she'd gain more weight, and feel the same vulnerability and rejection she'd felt when her boyfriend rejected her.

The protocol gave her the opportunity to recognize that without hunger, and by limiting her exercise, she would have to create happiness dependent on something else. The insight that she was addicted to exercise and emotional eating eventually disgusted her, and helped her redirect her motivation from losing weight, to creating and fulfilling her life without needing a crutch.

As her source of motivation changed from weight loss to emotional strength, she sought out opportunities to prove she could handle any situation. She went to bars, social events, and faced her biggest vulnerability—at night when she was alone.

Recognizing how her distress was linked to food motivated her to submit to, and to tolerate her feelings of emotional weakness. She chose to allow and accept her times of doubt and vulnerability, knowing that she needed to develop a sense of emotional self-esteem. She knew for her body to physically heal, she'd ultimately have to take responsibility for her emotional well-being without needing food. And she did.

What did we learn from this woman?

To me this is the most important lesson. To realize that self-esteem is just that: esteem fulfilled by one's self, not with the assistance of some other person, action, or object. This includes esteem received by physical accomplishment. Unfortunately, motivation to look good can cause serious problems for people who once defined themselves by over-eating. They have to change their definition to include a leaner, more socially accepted body. This can cause a massive fear of weight gain, and an extreme overemphasis on appearance. From binge eating, to bulimia, obsessive exercise, and sometimes, even anorexia—addiction goes from one extreme to the other.

In order to have freedom from feelings of emotional deprivation, it's very important with any weight loss that the individual separate his/her definition of personal value from his/her body, from needing restriction for control or from excessiveness. This prevents an ego dependent on physical accomplishment from experiencing the shame associated with fat gain, and allows the individual to rationally take care of his/her body without expecting, or being entitled to, an emotional reward in return.

In this case, the participant never fully gave up the identity she developed around competing in triathlons, but she did give up her desire to over eat which ultimately prevented further fat gain.

Like this participant, to start over you must expect to feel moments of vulnerability as you let go of what had been controlling your emotional well-being. Your desire for emotional strength must be stronger than your fear of emotional destitution.

Wouldn't you love to enjoy the carefree, child-like, emotional creativity not defined by eating, exercise, or dieting?

Until you are willing to independently create inner emotional abundance, you'll always be seeking outside of yourself to find it. You will avoid what makes you feel defective, compensating for your feelings of emotional poverty, and sabotaging any uncomfortable effort for independence. Developing intrinsic, emotional sustainability is the key to unlock the chains of emotional addiction, and slavery to food, dieting, or exercise. For this you must have tolerance, humility, and a desire to create happiness for yourself—independently, without needing food reinforcement.

Again, this brings us back to famine. Without the mass availability of food, emotional eating would *not* be an option. We would have no gluttony or dysfunction. Exercising excessively would be unwise, and weight would no longer be an important issue. Immediately you'd have to find creative and internal means to create emotional strength. Happiness would be a choice you create, and food would be a rationed miracle. Today, it's as simple as having a desire for emotional freedom, and making the choice to be self-sufficient—not because you're forced to by famine, *but because you want to.*

Chapter 3

Identifying Hunger:
Your Ticket to Emotional and Physical Freedom

To eat less, whether under forced rationing because of famine, or because you have a desire to eat less, you must understand the vital role hunger plays in guiding your body's primary need for food—as a hormonal stimulant.

Most people don't know that the fuel the body uses does not come from the food they just ate. Once you eat, a digestive process degrades the food, separates the vital nutrients from what's eventually removed from the body, and then it's stored into four different potential fuel sources:

1. Blood/cellular glucose,

2. Fat,

3. Glycogen, and

4. Body protein.

Of these four sources, we have no clue where the food will be stored, and when that fuel will be used. To make matters more complex, these four sources are not yet fuel. They are stored as "pre-fuel" or substrates that get converted into Adenosine Tri-Phosphate (ATP), or "fuel." ATP is to a human as gasoline is to a car. ATP is human gasoline.

The major difference is that ATP is not stored, but is immediately created from the breakdown of our stored "pre-fuel." Each of the four choices are special in that they provide ATP at different rates, different amounts, and uniquely, in different environments. For example, glycogen is a specialized fuel source that is converted into ATP very quickly, like for a fast demand, such as when a runner sprints. However, in this circumstance, there's only enough to fuel the body for about 30-to-60 seconds.

Body protein can also be used, but is obviously not the body's first choice, since using muscles and tissue for fuel would weaken the entire body. Body protein is only available when all other sources are not meeting the body's fueling demands. This is a typical symptom of starvation.

That leaves us with blood/cellular glucose and fat. Which one do you think the body prefers to use?

Most people answer "blood glucose," but are surprised to hear the body actually does *not* want to use fuel from blood glucose, unless it has to. This is because blood glucose is the only fueling source for the brain. Brain cells don't have the ability to break down their contents to create fuel, like body protein, fat, and glycogen do. The brain doesn't have "gasoline-making" abilities, and therefore, needs fuel delivered to it. This is great news because if your brain could use its cell contents for fuel, we'd have a problem with brain loss similar to what occurs with fat or muscle loss. This is why blood must be regulated to provide the proper amount of glucose—not too much, which is toxic, and not too little, which can cause death.

If the body were to compete with the brain for glucose, you'd have only enough to supply fuel to the body for about 40 minutes. It would be essential that you ate sugar every 40 minutes to preserve the brain from having to compete with the body to survive.

Separate from the brain's needs, blood/cellular glucose is not sufficiently stocked to fuel the body's needs, which leaves the tedious role of fueling the entire body to body fat. And fat just happens to be the most effective and well-stocked fuel source, to do the job effectively. Fat is the most effective source of ATP, as it can fuel the body of the average-sized human for about 30 days. In fact, fat is so effective at fueling the body with ATP that in a controlled environment, blood glucose could be completely reserved for just the brain and blood.

But to fully understand how the body preserves blood glucose for the brain when fat doesn't provide enough fuel, you must understand the hormones that influence the mechanism of starvation, and why the human body can live without food for a relatively long time.

Leptin's Role

When you compare all four potential fuel sources—blood/cellular glucose, fat, glycogen, and body protein—they all uniquely and circumstantially fuel the body in different amounts, and at different times. But, what determines which fuel source is used is highly linked to the hormone leptin. Leptin is a fat-derived hormone that was discovered in 1994. (See third section of this book for more details on this hormone.)

When elevated, leptin increases ATP production from fat cells. But if leptin levels decline, fat may not produce ATP at the rate the body needs it as fuel. If leptin is too low and fat can't supply adequate fuel, the body will tap into cellular and blood glucose.

As soon as the body begins to compete with the brain for glucose, mechanisms in the brain signal a defensive response. When leptin levels in the brain decline, in response to a rapid drop in blood glucose, enzymes become activated. These activated enzymes in the brain cause the physical irritation, agitation, and urgency to eat that our conscious mind feels as physical hunger.

This irritating signal to eat, caused from a drop in brain leptin levels, is for good reason—because food stimulates leptin.

When eaten, food in the mouth instantly stimulates the production of leptin, and as more food is eaten, each fat cell produces more leptin. As blood leptin levels rise significantly, fat can adequately break down to supply the body with ATP. In time, as blood glucose is restored to fuel the brain, and leptin levels increase, the enzymes in the brain that caused the urge to eat deactivate, and hunger subsides.

But what would happen if food were not eaten when the brain gives the signal of hunger?

Without food consumption to stimulate more leptin, glucose would quickly deplete. To prevent damage to the brain, there would be an increase in the breakdown of body protein (muscles and tissue), as well as stored glycogen. This shift in fueling, although weakening to the body, prevents life-threatening drops in blood glucose, and prioritizes fueling for the brain. The body responds with symptoms of starvation. Organs in the body that are stimulated by leptin weaken their signal (thyroid), and others that are suppressed by leptin activate (adrenals). Eventually this response establishes a new equilibrium that requires less fuel, matching what leptin is allowing fat to provide.

When you think about leptin, consider its vital role of allowing fat to supply adequate fuel to the body in order to prevent starvation. You need to eat when the brain gives you the green light that *is hunger*. However, if you eat without hunger, or eat after hunger subsides which is fullness, you may over-stimulate leptin, and this can cause toxic levels of fuel, which can harm responding organs.

Too much leptin influences the entire endocrine system: over-stimulates the thyroid, may cause hot flashes, sweating, adrenal fatigue, respiratory problems, sleep disturbance, decreases libido, etc. And worse, too much leptin can over-stimulate fat's production of ATP. Fortunately the body has mechanisms in place to prevent death from this toxic response caused from eating food without the hormonal need—and it is called fat gain.

When you combine excess ATP with an over-supply of leptin, you get brand new fat cells, and these cells are bigger and create more leptin than the other fat cells. Over time, with repeated bouts of too much leptin, more and more cells accumulate, and eventually you have fat gain in areas you never had fat before. With each new cell, your fat produces more and more leptin, and as the fat cells potentially get bigger, the amount of leptin you stimulate grows exponentially.

This growth in leptin production significantly reduces the amount of food you need to get the appropriate stimulus. This is like adding sound, not by increasing the volume to one speaker, but by adding more speakers; the sound becomes louder and louder as you add each new speaker. This magnification of sound is like the magnification of fuel from fat. If you have more fat cells, it takes less food to stimulate adequate leptin.

The more fat cells you have, the more magnified the hormonal response to food, which means you can live longer on less. This is a miracle! Fat gives you a great advantage if we had to ration because you would require significantly less food to reduce hunger, and you'd experience hunger less often. Think about it: how often do you experience *true* physical hunger?

Most people who have excessive fat experience hunger less often during the day. But when they do feel hungry, it comes on quickly and intensely. This may be due to a more rapid drop in blood glucose, when leptin finally declines in the afternoon. This rapid hunger may be caused by the incredible demand placed on blood glucose by this person's larger body size, and his or her need for more fuel than those who have smaller bodies.

People with less fat have more frequent hunger, and require more food to remove the hunger, but they don't usually experience such extreme sensations of hunger. The link between body fat and hunger shouldn't be surprising since we've known since the mid 1990s that leptin comes from fat cells, and that it reduces hunger.

It's really quite simple, the more fat cells you have, the more leptin you produce, which means you don't need as much food.

The downfall to this magnified leptin response to food is sensitivity to fat gain. Obviously, with more fat cells, you're going to get more leptin with less food, and if you eat more food, the risk is that your body will be forced to create more fat cells. Even worse, this can cause permanent damage to your other organs. This type of sensitivity does not happen to people who have less fat so they can eat more food without having a problem.

Unfortunately, we live in a society that encourages excessive eating and loves gluttony, which puts people who have more fat at a physical disadvantage. Their bodies require less food to adequately stimulate fat's fueling mechanism. But if eating less was common among your peers, then people who had more fat would be socially more acceptable as we'd be critical of those that needed to eat more food to relieve hunger. But let's get real. Eating less for most of us won't happen until something massive destroys our abundant supply and easy access to food. Don't wait for the social desire to eat excessively to go away. Instead choose to eat appropriately what your body needs, regardless of how much everyone else seems to think you need to eat.

The discovery of leptin and its role in fueling the body, completely conflicts with the outdated concept of monitoring calories from food. And to make counting calories even less effective, we have no clue what fuel source the food will eventually restock, and when your body will use it. Calorie counting is a concept we will look back at and laugh about. We'll wonder why we continued to spend billions and billions of dollars for diets to tell us when and how much to eat when no one has any idea how much fat we have, when we are hungry, or when food is needed hormonally. While the food guide pyramid changes as the USDA continues to learn more about nutrition and the body's fueling mechanisms controlled by hormones, the diet industry also continues to feed us information that may be causing even more and more fat gain.

In fact, predetermined meal times cause harm to people who have more fat. Pre-portioned meals also do. How does anyone know how many fat cells you have, how much leptin the food will stimulate, and, if the amount of leptin is adequate, relative to the demand for that fuel? Again, leaner people experience hunger more often, eat more food, and have less comparable damage to their entire endocrine system when they eat more potent foods, or eat too much than do people whose bodies contain more fat.

Eating-induced fat gain is associated to the death of hundreds of thousands of people. We seriously need to eat less. Most of our culture, including a huge number of children, is now dealing with fat-induced leptin sensitivity that decreases our need for food. In particular, we need to eat less of foods that powerfully stimulate more leptin, such as sugar and starch.

Candy, soda, and other sugar-based foods are highly toxic for people who have more fat cells. Not only does the sugar immediately get absorbed into the blood stream, but it also stimulates more leptin from fat cells than other food types. This means that sugar can cause toxic increases in blood glucose, magnifying the body's fueling mechanism from fat even more—causing rapid and more substantial fat gain, and even further damage to other organs, which can lead to early death.

When you look at diseases associated with obesity, it's easy to blame the fat, but now it's time to reconcile with the body. We need to recognize that our behavior with food has abused our built-in system of hormonal feedback, and that hunger has been ignored.

**Whether you have two hundred or ten pounds of fat you
need to lose, to eat functionally you have to listen to hunger,
eat when it's present, and stop when it subsides.**

When hunger increases, your leptin levels are declining, and food, as a stimulant, is needed. As hunger subsides and leptin levels increase, food consumption is no longer appropriate. To follow hunger as a guide, you must be keenly aware how foods differently influence the speed and longevity of this signal. And you can't compare—when, what, or how much others eat—to your body's needs. People who have no idea what real hunger feels like, find this difficult to understand, especially those who've lived their entire lives eating dysfunctionally by the artificial dictation of diets or unrestricted gluttony.

Again this brings us back to famine, using the physical feeling of hunger as a rationing guide, and trusting that your body knows exactly how much it needs, and when. You can't compare your body's rhythm of hunger to anyone else's, since different types of food combined with the number of fat cells you have, greatly influence the timing and need for food. Individually, you need to recognize hunger, and stop comparing it to the limitless reasoning that is a normal part of our dysfunctional eating culture. This is like creating your own famine, not because you're forced to eat less, but because you want to. But for this to work, you have to know what true physical hunger feels like, apart from emotional need.

**You can't compare your body's rhythm of hunger to anyone else,
as different types of food combined with how many fat cells you
have, greatly influence the timing and need for food hormonally.**

Chapter 4

Eating Less, Because You Want to—in a Culture That Loves Gluttony

What would happen if eating less were "cool"? This would drastically reduce the demand for food, completely revolutionizing the food industry, and literally eliminating the need for the weight-loss industry.

- What would happen to associated illness and disease with significantly reduced obesity?

- How would this influence the pharmaceutical industry, including the FDA?

- And to make matters even more radical, how would all of these changes affect our country politically?

Wow! A drastic revolution in our current food-laden cultural model would have major implications—favoring our emotional and physical well-being. Still, this type of change requires our culture, as a whole, to redefine what is socially acceptable, and as individuals, to understand that we're emotionally capable of eating significantly less, even when we could easily have more.

A permanent solution requires finding new ways to fulfill emotional needs, without creating a new dysfunctional addiction. It also calls for redefining a new normal way of eating—a way that doesn't predispose fat gain, and doesn't require you to weigh yourself for motivation.

Start by looking at the role eating plays in your life.

- When do you eat without hunger?

- How often do you eat until you're full?

- How do you justify over-eating or eating without hunger?

- When do you eat because you feel vulnerable or emotionally weak?

- Do you even know what hunger feels like?

- Could you stop eating when hunger goes away, no matter how fabulous the food tastes?

Rethink how you eat, and imagine what it would be like to minimize food intake—as if you are forced to by famine, and the consequent rationing.

Rationing

Visualize how you would eat to conserve enough food to feed yourself and your family for as long as possible.

- How would you ration the food daily?

- What reason would make eating appropriate?

- Would you eat until you felt uncomfortably full?

- Would you eat without hunger?

- Would you force yourself and others to clean your plates?

- Would you let yourself get so hungry you'd irrationally eat more than you needed?

In an ideal world, you'd only eat when there was some hunger, but not too much. Then you'd eat only until that hunger was eased, but not a bite more.

What would you do when bored? How would you celebrate if there wasn't food? Would your motivation to eat less come from monitoring your weight? When I ask my clients these questions, the answers are predictable. Most feel they'd be grateful as long as they had enough food to keep their family alive. They almost always point to hunger as a guiding sense about when to eat and how to instinctively avoid over-eating.

With a change in perspective, suddenly eating less doesn't feel punishing at all, and weighing yourself as motivation, seems stupid. This view changes what is perceived as emotional hardship—away from the superficiality of weight and emotional eating, to the profound and meaningful importance of life and family. Food instantly is viewed only as a physical necessity and emotional eating is obviously viewed as irrational.

Visualizing this scenario forces you to independently assess your emotional strength, as if eating emotionally were NOT an option. For most, a sense of emotional strength and confidence is instantaneous when they have no other option.

The question we need to answer is: Do we really need mass destruction and famine to force us and our entire culture to reprioritize the role of food, and to functionally eat less?

The problem is that most people actually like emotional eating, and they won't give it up unless they are forced. If you choose to eat less as a new norm, expect to be offered food when you don't need it at all, and—to be served more than you need. To make understanding this easier, it's imperative you actually know what hunger feels like so you can differentiate physical need from emotional desire.

Physical Need, or Emotional Desire?

Differentiating is not as easy as it seems. Physical and emotional need for food can become so intertwined that detaching one from the other can be difficult, and recognizing the difference, almost impossible. To make matters more difficult, years and years of dieting, as well as attaching judgment to food, can make people feel incapable of deciding for themselves what and how much to eat since they've learned to rely on others, like the weight-loss industry, for that information. Not only can this make people mistrust their bodies, and the food they eat, but also they feel insecure without the restrictions of a judgmental diet.

Relearning to trust the physical cues of hunger requires that you let go of calorie counting, pre-portion controlled foods, and pre-determined meal times. It demands that you end all emotional judgment about food. You must completely start over with your diet so you can develop confidence in your ability to listen to your body when eating—without the extreme opposites of our cultural excess, and without a diet's control.

The goal is to recognize how your body gives feedback, and to allow this feedback to guide when and how much food is needed, since varying foods influence this system differently. After you've mastered functional eating, the next goal would be to apply nutritional intelligence to the foods you choose. To be successful, you need to focus on eating functionally before you attempt to learn to eat nutritionally. This requires you to trust and learn to use your sense of hunger. This is where the protocol comes in handy, and why its profound value surpasses the superficiality of weight loss alone.

Case study

One of the most difficult situations occurs when food and eating replaces one's need for love. One of my male participants spent most of his money and extra time thinking about food, planning his next meal, and then eating when he had free time during the day. Food was his hobby, his entertainment, and his reward—and it started at a very early age. As other boys were playing sports and chasing girls, he spent his time at home—eating. During high school he reached 300 pounds, and eventually over 400 pounds during college.

He'd dieted many times, losing weight with every short attempt. However, as soon as the diet was over, he resumed his love affair with food. Without eating, he felt lonely and sad. Dieting took away everything—his activities, his hobby, and his life—because it all revolved around food. Eventually, he gave up on dieting, and began to use his large size as a source of ego.

The larger he got, the more he used his size to make jokes, to differentiate and distinguish himself from others. He eventually took pride in his size and developed an identity around his massive obesity, as well as his love for food. He even made fun of others for exercising or for eating less. However, his size came between him and having a loving relationship. This realization prompted him to do something—and he chose the hCG protocol.

His motivation was so strong he was willing to redefine himself entirely—from the way he looked, to his life with food. His first protocol was easy. Since he prided himself on his ability to take on this challenge he didn't cheat, and by the end, had lost over 50 pounds in 40 days. He continued to lose during his maintenance, and before starting his second protocol, was down over 70 pounds. However, his second protocol was a different story.

He wasn't as miserable with his weight, and he didn't have the same emotional distaste for his resulting fat. He started cheating within the first ten days, and was amazed at the fat gain he experienced with such small cheats, a bit of something here, a drink of alcohol there. As he came closer to the end of the protocol, he'd only lost twelve pounds after six weeks on the very low-calorie protocol. He was so frustrated that he started to blame the protocol for not working. He argued that with other diets, he could cheat liberally without conflicting with results. Although he still lost two pounds of fat a week, he felt deprived by the constraints, even though physically he felt significantly better.

He struggled during the maintenance phase as his desire to eat emotionally increased, and his urgency to lose weight decreased. He assumed his third protocol would make up for what he was gaining back, and let go of any constraint when eating.

By the time he was ready to start his third protocol, he was embarrassed by the weight he'd regained. His third protocol went well, cheating minimally and getting his total losses past the 100-pound mark. But still, the amount of fat he lost didn't make up for his emotional desire to eat. Each time he'd transition into the maintenance phase, he'd struggle. Without rigid boundaries, he'd find himself eating when he wasn't hungry all the time.

He felt without strict rules he was vulnerable, and without motivation to lose weight, he had no reason *not to indulge*. Each time he'd let go of all logical restraint and gain weight during the maintenance breaks, then he'd work hard emotionally to restrict during the protocol. It was a feast before famine cycle.

What can we learn from this man's story?

This participant never allowed his body sufficient time to lose enough fat to completely change the way he looked. His persona had developed such a dependence on being obese that he couldn't find the desire or reason to continue to lose weight. Despite the physical damage to his body with the excess weight, losing any more fat would require that he redefine who he was without an obese body, and without eating as his primary focus in life. To him, changing his body felt like *the death of who he thought he was*. The idea of living differently was too scary, made him too vulnerable, and he felt too weak inside to experience anything other than what he already knew.

His father was an alcoholic and his mother emotionally unavailable, so he'd often been left to provide for himself, and food was that provision. So when he was challenged to divorce this relationship with food, it forced him to question if he knew how, and if he was capable of finding love elsewhere. Was he willing to open himself to others? To feel vulnerable to the possibility of rejection? Today, he is engaged to be married, but has yet to finish his quest to redefine himself as a smaller person. He's kept his weight off but has over 100 pounds to go.

Chapter 5

The hCG Protocol: Emotional and Physical Rehab

The hCG protocol provides the perfect environment for rehabilitation from emotional eating. Because hCG may significantly influence the mechanisms that regulate hunger, the more fat a participant has, the less physical hunger he experiences.

HCG magnifies the role leptin plays in regulating fat metabolism, which significantly replaces your hormonal need for food. The combination of needing less food and having less hunger makes our normal way of eating completely inappropriate. This change requires that the participants live and be exposed to life without the emotional comforts of food—otherwise fat gain and hormonal problems could be the result. This creates an amazing environment to get rid of our harmful and physically destructive old patterns in order to recreate a new functional way to eat that promotes health and well-being.

By unplugging your life with food and rebooting it, you have a chance to start over with a lifestyle that doesn't cause fat gain. Habits that made sense before, no longer apply. Eating at mealtimes, for entertainment, or when bored, can cause immediate problems. The protocol environment not only magnifies the physical response to eating, but also puts dysfunctional motivation to eat under a microscope. All reasons to eat that have nothing to do with hunger become obvious as the protocol constraints are clear, and cheating is apparent.

It's a common belief that with minimal hunger, the protocol will be easy. But for most, hunger is not why they eat or why they cheat on the protocol. Most people eat out of habit, and, without hunger, those habits are immediately questionable. But the habitual desire to eat is less challenging during the protocol than the emotional desire to eat. The emotional desire to eat creates the most obvious roadblocks, the most turmoil, and causes the protocol to feel depriving and difficult.

Just because you have minimal physical hunger, doesn't mean there isn't maximum emotional hunger, and this is what complicates the protocol for most participants. They may cheat in three ways: by eating food that is not approved, by eating more than the maximal limit, or by eating without hunger. Emotional deprivation is usually the reason behind any cheating. The strict protocol and hunger-free environment amplifies emotional insecurities because cheating exposes the desire to eat as an emotional crutch.

Emotional deprivation on the protocol magnifies the dysfunctional rationale to eat that most people completely ignore. For example, it's common for participants to reward themselves with a cheat after a long day. They eat in the evening, and justify the cheat because they're only eating a small amount of something.

One of my clients cheated with a spoonful of peanut butter as a reward after a long day at work. Since she recognized she wasn't hungry, she was forced to acknowledge her dysfunctional relationship with food. The following night she tried again to follow the protocol, but this time she felt the need for a reward—and she chose to eat spinach. Spinach is an accepted food item on the protocol, and wouldn't cause a significant hormonal response, however she was *not* hungry. Eating without hunger, whether it's with food that is or is *not* on the protocol, is an indication of a behavioral problem.

In this case, the woman's behavioral problem was eating to fulfill herself emotionally, even though she knew, and was informed that cheating would was cause hormonal imbalance, and as a consequence, fat gain. By becoming aware of her behavior around hunger, the reason to eat for emotional gratification became very clear, as if we were watching her behavior under a microscope.

The ability to distinguish true hunger from emotional hunger allowed this participant to see a wider perspective of her behavior with food, and how she went about justifying her need to gratify her emotional deprivation. She wasn't a binge eater, and didn't eat gluttonously, but she'd eat without hunger all the time. Eating was how she'd console herself after a hard day's work. Without food, she felt lonely and agitated at night—as if she had worked hard, without reward.

She recognized she ate at night to shift her mind away from work, and to "decompress." Eating helped her focus on something that felt less draining, and more rewarding. And because her hunger was minimal during the very low-calorie protocol, she was forced to recognize that justifying eating for these reasons was completely dysfunctional. Her desire to eat had nothing to do with hunger but was caused by her discontent with her current job, and her need to do something she actually liked. For her, eating was a passive way to deal with her unsatisfactory work situation that required little effort, and to receive something back emotionally after a long day of unrewarding work and emotional deprivation.

As she continued the protocol, she proactively focused her efforts on creating emotional fulfillment, which immediately removed her passive desire to eat. All of this happened because she could finally recognize she truly wasn't hungry at night.

This example is typical of many participants, but their discontent may not stem from a work situation, but from their lack of control of something or someone in their lives. I've had some participants quit their jobs, change their religious beliefs, or end a relationship because they realized they ate to mask the emotional hardship caused by what they didn't intrinsically believe, like or want. And all along, they wrongly thought their obesity was the problem.

For others, the need to eat isn't caused from discontent. Eating excessively can be a learned behavior taught and enforced by their parents' model. Many are taught to eat all the time, for no reason at all. Some people rarely feel true hunger because they are taught to eat as soon as fullness is gone. They are taught to eat three meals, regardless of hunger or the lack thereof. They tolerate severe hunger while they wait until it's the proper time to eat, and when it's finally time, they irrationally eat everything until it's gone to make up for the next long waiting period they know is coming.

In social settings, the majority of us eat dysfunctionally or excessively. We eat before we do something fun, after the fun is over, and during the process. If you meet friends and family for fun, it probably entails eating food, and lots of it. What would it be like to go out to a movie with friends and not eat popcorn because you're not hungry? Could you go to a cocktail party and have fun, without drinking or tasting the food? Many participants feel punished and deprived when they socialize while they are doing the protocol because they're being challenged to have fun without eating.

Again, the challenge for most on the protocol is not caused by physical deprivation, but by the emotional deprivation they create by not being the proactive source of their own emotional fulfillment. They think fun and happiness must be given to them, rather than seeing emotional satisfaction as a feeling they must essentially create for themselves. Therefore the protocol is a punishment— it takes away their fun. This is why monitoring weight while dieting is a very important emotional reinforcement for people who have emotional eating addictions.

These participants tend to complain more, weigh themselves obsessively, and find every day of the protocol a struggle, even when they achieve incredible physical benefits that surpass the superficiality of weight. They usually profess that the protocol isn't realistic—and it doesn't work. But because they justify cheating, and are biased toward their emotional needs, they don't recognize their addiction to food as a way to have emotional fulfillment conflicts with how the protocol works.

Even with a significant amount of weight loss during the protocol, you can't expect to maintain any fat loss if you're not willing to change the way you eat normally. And no amount of fat loss will cure an emotional problem. For that, we know of no protocol—and no miracle pill.

No one but yourself can create the motivation to change your relationship with food. Only you have the power to create an intrinsic desire to eat less than what is available. Despite the gluttonous freedom to eat without limits, and the ability to eat for emotional reasons, you have to want to be accountable for your own emotional abundance.

The protocol is a miracle, but not for your emotional addiction. Only you can change the emotional role that eating plays in your life. You must be prepared for desires to eat that have nothing to do with hunger, and you must want to find a new relationship with food that isn't emotional.

If you aren't hungry, the questions you need to ask yourself are:

- Why would you need to eat?

- If there was famine, and the choice to over-eat wasn't an option, would you feel emotionally deprived?

- What are you receiving from food that you feel unworthy or incapable of creating emotionally?

To find out, you need to first know what hunger feels like, and from there, be willing to live as the creator of your emotions, rather than expecting food to supply them. This is where the hCG protocol can come in handy.

You can use the protocol to redefine the role of food for yourself. It provides a blank slate, a "do-over," and a means for you to redefine a new "normal." The protocol provides a famine that you can use as an opportunity to relearn the true nature of hunger. With this, the true enemy in your battle with weight is exposed, not as fat, but as your desire to over-eat, and to eat emotionally.

With emotional freedom from eating as the ultimate goal, hunger becomes the most valuable guiding mechanism that helps you completely detach emotion from the choice to eat. And when the participant allows the body to control when and how much food is eaten, he gives his body time to heal, permanently altering his relationship with food.

However, if your effort isn't focused towards using hunger as a guide to decipher emotional need from physical need while on the protocol, you won't understand why you emotionally need food. You won't change the way you feel about emotional eating, and you'll go back to the behavioral problems that caused your weight gain in the first place.

Food portions are still going to be too big, and social eating will always be excessive as a normal part of our culture.

You have to want to eat less in a culture that loves and encourages gluttony, without needing weight loss as a reward. To do this, you need to know what hunger feels like, and then do whatever your body tells you. You become the observer, and your body—the controller.

Take the culture's disapproval of body fat and emphasis on weight, and redirect that effort towards changing your behavior. When you're challenged to find emotional security, you're forced to create it without needing to eat. Like taking the pacifier away from a baby, you have to believe you'll be fine without it.

If after the protocol, you're willing to let your body control when and how much you eat, no matter how vulnerable you feel, you've won the battle—physically and emotionally.

Chapter 6

The hCG Protocol: Another Fad Diet?

Until both the businesses that provide hCG, and the participants that do the protocol, discover and confirm the protocol's true potential, it will never gain credibility as a form of therapy. It will be another fad diet.

How much of the American population has participated in multiple fad diets? The answer: Enough people diet often enough for it to be a multibillion-dollar industry.

In fact, if you were battling weight in the 1970s, you might have already done the hCG protocol when it was called "the Pregnant Woman's Urine Diet." You'd go to the medical clinic every day for your injection, and you were asked to follow a strict diet, similar to the one spelled out in Dr. Simeons' *Pounds & Inches.* You lost weight and you gained it back. Why? Because no diet will ever work for problems that are caused by an emotional problem.

Case Study

One of my protocol clients was the wife of an executive, whose role required they attend social engagements, often with people she didn't know. To ease her fears and feelings of insecurity, she'd drink wine to feel more secure, and eat to avoid conversations all together. During her first protocol, she avoided these engagements because she had very little confidence in her ability to cope with the social anxiety. Added to that, she also had a very stressful relationship with her mother-in-law. She often felt angry and irritable, eating to get emotional relief.

Although she struggled both on and off the protocol, her weight loss was an amazing motivator, and she followed the protocol meticulously.

During the protocol, each day she'd weigh herself, and was rewarded with a loss of over thirty pounds, and a mass amount of body size in less than six weeks. As she approached the end of her protocol, her desire for more weight loss diminished, and her excitement to eat without the restraint became her focus. She was so proud she'd gone through her first protocol without cheating that she mentally planned to celebrate, and salivated at the idea that she was done with her committed time of restriction.

The first day into the second phase of the protocol everything she learned about the value of hunger and avoiding fullness went out the door. A week after her last day of hCG injections, she called in serious pain. At first she thought she had the flu, but when the stomach pain continued, she finally decided to call. Even though she was warned to add fat back into her diet slowly, she was so excited to get back to her normal delights and emotional eating without limitation, she did not follow all warnings and directives. She immediately indulged in sausage, bacon, creams, and all the things that she felt the protocol took from her. Within a few weeks, she gained all of her weight back.

She also continued to eat and drink to mask her social insecurity and her weakening ability to handle stress. Her excuse: "I can always do another protocol." She feasted before every famine, never really gaining the emotional strength necessary to end her dependence on food and wine to tolerate the normal stresses in life. She constantly blamed her body and eventually also blamed the protocol.

Many people cyclically do the protocol because they know how easy, effective, and fast it is to lose weight. The immediate gratification of weight loss, and less time away from their emotional addiction to eating and drinking, makes it very appealing for people who hate the physical consequences of fat gain.

What can we learn from this woman's story?

She did *not* understand the role her emotions played in weight gain and loss. She did not know how to deal with daily normal, as well as highly stressful, events without turning to food and drink. She did not understand and she was unwilling to accept that she needed to follow the protocol's strict guidelines, to prevent the weight from returning, and in her case, from her getting sick because of the sudden influx of rich foods into her diet too fast and copiously.

If instead she had removed the scale and her motivation to lose weight, she would have been forced to find a different reason to choose not to eat or drink when she experienced social anxiety and stress induced by outside factors. Once she recognized her need to eat and drink, she approached her next protocol with a different mindset.

She chose to only weigh once a week, and went to social functions with the mindset that she would engage socially—even though she was fearful. Although she could not change her mother-in-law, she decided that she'd have to change the way she perceived their relationship, not depending on food to make it all better.

After her second protocol, she lost everything she'd regained after her first protocol, and more. But the real benefit was that she felt more confidence in her ability to tolerate stress.

Success stories like this show that the value of the protocol is more profound than just the superficiality of measuring how much weight is lost. It has the potential to be more than just another fad diet.

Unfortunately because most participants, skeptics, doctors, and even the FDA, are not adequately educated to understand starvation, or how hCG influences leptin, we have yet to have an intelligent debate about the value and the hormonal influence of this protocol. Until then it will remain less available for people who desperately need it.

Skepticism is good, but when the use of hCG for weight loss is disputed by outdated concepts of calories or research based on subjective assessments of perceived hunger, observations of scale weight, and circumference measurements, this shows a lack of current information concerning the hormone hCG and poor judgment. The FDA appears archaic, continuing to refer to old, insufficient research as proof the hCG doesn't influence the body during a starvation diet. They're obviously not up to date with the science of starvation and leptin, which directly influences a person's hunger and their hormonal need for food. Otherwise the FDA would suggest laboratory investigation that meets today's standards, appropriately applying the protocol from a hormonal perspective.

Those who are convinced the protocol works hormonally, need to seek out legitimate answers from those that question hCG as a hormonal influence on the body. Until physicians, practitioners, business owners, instructors, and participants move past outdated material to seek answers that make scientific sense, Dr. Simeons' protocol will never gain scientific credibility.

Recently, as consumers have bypassed the FDA's insufficient argument, they've let go of their need for the FDA to control their decisions, and are demanding hCG anyway. The protocol has made an explosive resurgence, this time at the demand, and often the control, of the consumer.

The hCG protocol started its recent resurgence on the Internet because local doctors were reluctant to prescribe such a drastic starvation diet. But now more people are seeking medical advice and attention during the process. In the last five years, there has been an explosion of medical hCG programs being implemented in weight-loss clinics, medi-spas, chiropractic offices and wellness centers. Unfortunately, speculation and controversy exist because any new science applied to Simeons' protocol is lagging far behind the demand. Both the medical community and the hCG weight-loss industry continue to argue for and against different bodies of outdated research. This research is based on observations that don't scientifically prove or disprove the physiology behind Dr. Simeons' explanation of why his protocol works.

The downfall to Dr. Simeons' explanation was his emphasis on observations to prove his theory about the metabolic process. He knew his observations fell short of laboratory experimentation as he quoted in the foreword of *Pounds & Inches*:

> "...I shall be unashamedly authoritative and avoid all the hedging and tentativeness with which it is customary to express new scientific concepts grown out of clinical experience and not as yet confirmed by clear-cut laboratory experiment. Thus when I make what reads like a factual statement, the professional reader may have to translate into: clinical experience seems to suggest that such and such a working hypothesis, requiring a vast amount of further research before the hypothesis can be considered a valid theory. If from the outset establish this as a mutually accepted convention, I hope to avoid being accused of speculative exuberance."

Dr. Simeons knew he hadn't clinically validated his hypothesis, but he hoped his observations would spawn laboratory research. Without applying the obvious clinical research necessary to further understand his ideas, no valid answer existed to either prove or disprove his theories. Instead, refuting research concluded the protocol's ineffectiveness by disproving his observations *without considering his hypothesis*.

Researched in the 1960s to 1970s, most of the double-blind studies found that:

1) hCG didn't reduce appetite any more than the placebo group,

2) hCG participants didn't have faster weight loss, and

3) the protocol didn't redistribute fat differently.

However, these observations do not address Simeons' most important beliefs: hCG's affects on the brain, and how it influences metabolism when combined with the protocol.

Instead, the studies tested hCG as if it was a weight-loss stimulant, which is very different from testing the conditions of the protocol and how it manipulates the metabolic system of the body. To base the hCG protocol's effectiveness solely on where, how much, and how fast fat is lost, falls short of what is necessary to validate his beliefs. Unfortunately, arguments both opposing and supporting Simeons' observations, point all attention towards the outdated research. No wonder the medical community has been slow to accept the protocol.

The most recent report, "The effect of human chorionic gonadotropin (hCG) in the treatment of obesity by means of the Simeons therapy: a criteria-based meta-analysis," was written in 1995 by Lijeseen, Theeuwen, Assendelft, and Van Der Wal. They concluded:

> "...there is no scientific evidence that hCG causes weight-loss, a redistribution of fat, staves off hunger or induces a feeling of well-being. Therefore, the use of hCG should be regarded as an inappropriate therapy for weight reduction, particularly because hCG is obtained from the urine of pregnant women who donate their urine idealistically in the belief that it will be used to treat an entirely different condition, namely infertility."

These authors were surprised by the lack of research after 1977, and found that the small amount of research they reviewed was lacking in methodological quality. [1]

Again, this summary determined Simeons' hCG protocol was inappropriate as a therapy for weight loss. Using observation, rather than any evidence of the physiological process produced by using the protocol. In particular, Simeons' theory of metabolism was never tested, and no consideration was given to testing physical response to the protocol's first- and second-phase constraints, with or without hCG. Dr. Simeons, and all of the old research including the 1995 summary, didn't have today's evolved understanding of energy homeostasis and of starvation.

Today, appropriate laboratory research would observe blood glucose levels, insulin levels, leptin levels, thyroid function, adrenal response, resting energy expenditure, cellular mitochondrial adaptations, respiratory quotient, and much more. To relevantly conclude whether hCG affects the hypothalamus, hunger, fat, and metabolism as Simeons theorized, the emphasis would be on explaining the physiological process of the protocol, rather than justifying its use or disuse, based on comparing subjective results.

It's interesting to examine the arguments about hCG, not only from those who outright reject the protocol, but also among those within the hCG community. For example, hCG can now be administered either through the original injection method, orally, or topically. Because of costs and inconvenience, we've seen a mass movement of participants using the homeopathic application of hCG.

The popularity of homeopathic applications, however, only proves how easy homeopathic hCG is to get, and suggests how desperate we are to lose weight.

The weight-loss industry preys upon this desperation, and is marketing the protocol like another fad diet. The bias against any drastic diet is, in part, due to the fact that the weight-loss industry is a failure due to the short-lived and half-hearted participation of its consumers.

No fad diet, or *any* diet, can overcome personal behavior. When we see hCG advertised, we see unrealistic statements such as, "Lose belly fat without exercise." This type of mentality has caused the hCG protocol to be added to a long list of fad diets, and makes it look like all the rest. Achieving weight loss without accountability is unrealistic, no matter how amazing the diet. Integrity is immediately questionable because of the overemphasis on how fast fat is lost. Making matters worse is the overwhelming and confusing influx of products and unsubstantiated claims.

In addition to oral drops, topical creams, dissolvable pills, nose sprays, skin patches, and injections, businesses are selling pre-protocol cleanse kits, B12, coconut oil, appetite suppressants, stimulants, and anything else that will make a quick buck for the suppliers. Blogs, websites, groups, and all sorts of "expert" advice, are perpetuating misinformation, naively tampering with the protocol, and falsely treating the protocol like any loosely followed diet.

I don't think that all people advertising and selling the protocol intend to make it a hoax. However, when it's treated with the same quick-fix marketing and taught with the same fad mindset, the result for the participant will be the same: temporary. Therefore, with medical professionals hesitant to prescribe, consumers have bypassed relevant science to experiment with an easily accessed but misunderstood protocol, making it even less credible.

If a proven explanation for how it works existed, it wouldn't be considered a fly-by-night diet scam, and there would be less debate within the hCG community. Without standards or integrity in much of the Internet-based hCG weight-loss industry, you should be skeptical. But it would be more appropriate to be skeptical about the integrity of those who provide the hCG, and the adherence of those who attempt the protocol, rather than the effectiveness of the protocol.

Most Internet-based hCG businesses prey on the fact that we are desperate to lose fat and will spend money on anything that promises fast results. And because participants fall for the bait, and attempt the protocol with short-sided motivation and misguided information, the protocol looks no different than the other fad diets. Many clients have asked if I guarantee results, and the obvious answer is no because I can't trust that the participant will actually follow the protocol restrictions. It's not that the protocol doesn't work, but rather, the participant doesn't specifically and diligently follow the protocol.

Using Today's Science to Prove the Protocol's Worth

Putting all past research and theory aside, and with the utmost gratitude and respect to Dr. Simeons, we need to make an effort not to prove or disprove Simeons' observations, but to understand how the protocol works. Using today's modern science, we can formulate a new hypothesis, a new explanation, and find a better way to forewarn participants that the protocol is used *not* as a diet, but rather as a hormonal therapy.

The protocol will not work if we continue to apply it like a diet. Even a simple taste of food may cause an excessive hormonal response. This response is *not* typical of diets. Cheating and misdirected efforts significantly influence a participant's hormonal system, causing fat gain and other hormonal-induced problems associated with obesity.

Without proper instruction, participants may further harm their physical well-being, and without proper motivation, they could make their emotional eating problems much more addictive. And, even if the protocol is strictly followed, the participant must change the way he or she eats normally, in order to maintain any amount of fat loss.

This requires the participant change the role that food plays in his or her life, and the protocol cannot do that. Until we are motivated and willing to change our normal way of eating, we will always "battle" our weight.

Chapter 7

Conclusion: Who Are You Without Obesity and a Diet?

It's important that participants understand their identity—how they see themselves—especially if their identity is defined by eating, dieting, and/or size. No matter how much you think you hate being overweight or obese, if you aren't willing to recognize your identity apart from the act of eating, dieting, or the way you look, any attempt at something different will fail.

To create motivation to change, you must first recognize your false definitions of yourself, and be willing to seek something different. But this requires an incredible amount of humility to seek out, and then allow feelings of vulnerability, unpredictability, and instability as you intentionally seek emotional strength without food as your crutch.

This experience forces you to recognize yourself from a new perspective; to see that your authentic value isn't made better or worse by the way your body looks.

For those who are anorexic, can you relinquish control and eat, even though it will cause inevitable fat gain? Can you love yourself anyway?

If you're obese, can you allow the emotional deprivation that occurs without eating, to help you begin to identify yourself in a slender body? Would a smaller body change how you've defined yourself?

A clear identity of who you are—not defined by your body, behavior, looks, and things dictated by others—is a fundamental part of self-esteem. Self-esteem is tested when you give up emotional eating. With this added dimension, the protocol takes on a new meaning and a new way to think about choosing to eat less.

If you are NOT willing to recognize your identity separate from the act of eating, dieting, or the way you look, any attempt at something different WILL fail.

Case Study

Some people have no idea that they eat emotionally, so they expect to follow the protocol perfectly and are surprised when they cheat.

One client had lost over fifty pounds, exercising daily and changing his diet. His weight had stabilized, and he found he was getting lazy and complacent with his diet. His motivation to do the protocol came when he became determined to lose the last extra thirty pounds to achieve his ideal weight and improve his overall health.

Thinking that with reduced hunger and immediate results following the protocol would be easy, he was not prepared for the little emotional triggers that would tempt him to cheat during the process. He cheated when food was offered, in order not to offend the person who offered it. He wanted to eat to cope with relationships he didn't like—and were not good for him. A cheat here and a cheat there, made him realize he had more emotional desires to eat than he realized or anticipated.

Cheating helped him to identify his hidden emotional insecurities; and he found he ate to compensate for discontentment and daily irritations. He decided to let go of relationships that weren't healthy, and began to listen to his body with a more keen discernment of true hunger. By the last three weeks of his very low-calorie protocol, he'd stopped cheating, and eventually reached his ideal weight. But the most gratifying part for him was becoming aware of his emotional conflicts, previously masked by food, allowed him to resolve issues and recognize that eating doesn't emotionally improve anything.

Not everyone struggles through the protocol. In fact I have more success stories than failures. One of the most interesting observations I've made is that in general, women who've been through menopause tend to have the best long-term results. Why is that?

- They are, by and large, motivated more by health, and the desire to feel good.

- They tend not to obsess over their weight, which reduces emotional hardship throughout the process.

- They seem to be more emotionally secure, are comfortable with who they are, and less motivated by superficial goals created by looks and sexual appeal.

Unfortunately many women who've lived most of their lives eating food functionally (not because of emotional prompts) can end up having weight issues after menopause—and women who have had weight problems, have even more later in life.

During menopause, ovaries under-stimulate hormones that consequently stimulate leptin, decreasing the hormonal need for food at random and unpredictable times. How they've eaten for years is no longer appropriate, even if they exercise more and eat less. Women tend to gain new fat cells that collect in the central or visceral parts of the body, and the more fat cells they gain, the more leptin they produce. This, in turn, reduces their need for food even more. They can't eat the way they used to, and they have new symptoms and new hormonal problems.

After menopause, many women never feel the same. With the extra gained fat, they experience symptoms of hormonal problems that weren't there before, such as adrenal fatigue. Too much leptin reduces adrenal stimulus from the anterior pituitary in the brain, which can cause issues with mood, energy, motivation, and vigor. Women tend to feel drained, exhausted, and tired without having done the work or exercise to merit that type of physical fatigue.

Every pound of fat added during menopause, adds exponentially more leptin, and if eating habits remain the same as before menopause, symptoms of adrenal fatigue could result. The solution: minimize any cause associated with eating without hunger, and also remove the excess fat that causes the hormonal hypersensitivity.

Women who don't eat emotionally, who normally eat functionally, find the protocol a miracle cure for hormonal changes caused by menopause. The abdominal fat that changed their figure goes away, and the tendency towards fat gain is removed. Even better, hot flashes go away, sex drive increases, sleep improves, and women return to the healthy status they had previously.

Many of these women don't have much weight to lose. They ate functionally before menopause, and they gained less fat than most. I've had some post-menopausal women come in with only fifteen pounds to lose. Their biggest complaint with the protocol was having to force-feed during the first three gorging days (explained in section 3). The very low-calorie protocol is easy, they love the simplicity, and they usually feel great all the way through. They're amazed at how quickly their figure returns, how much more energy they feel, and how grateful they are for the protocol.

The protocol isn't hard for individuals whose relationship with food isn't obscured by emotional eating. For many who've gained fat due to hormonal complications, the protocol can be the easiest form of hormonal therapy—and when combined with continued functional eating, could prevent any further hormonal symptoms and need for medication.

But even with emotional hardship, I've seen the protocol work for people with some of the most complicated and lifelong emotional battles.

Case Study

One woman's success particularly stands out. I remember the day I met this woman I'll call Ann. She was sixty-eight years old; 5'2 inches tall, and weighed over 230 pounds. A friend referred her to me after a doctor informed her she needed a total hip replacement—and she needed to lose a significant amount of weight before she could have it. As she considered surgery to reduce her stomach size, she also looked into the hCG protocol. She decided she'd rather self-impose the extreme caloric restriction through the protocol, rather than go through surgery to accomplish the same thing. Less risk, less cost, and ultimately, lower risk of death. But she knew in order for her to relieve her chronic hip pain, she'd have to lose all of her excess weight—and keep if off.

Before I consider participants for the protocol, I seek a thorough understanding of their relationship with food. Ann seemed to be ready for the emotional challenge.

But I was definitely concerned. For over 20 years she owned a bakery with her husband, a well-known chef, and it was her job to make and decorate cakes, and desserts. On top of that, she used food emotionally, starting at a very early age, and had spent the first half of her life drinking alcohol as an emotional crutch. She'd been sober for over 20 years, but had continued her desire to eat without limitation.

What convinced me she was ready for the protocol was that she'd already spent eighteen months exercising at least five days a week, reducing her food intake, and meeting with a dietitian. She had dramatically decreased her emotional reliance on food. Her frustration and desire for something more drastic came because her weight hadn't dropped in over six months, even with increased activity and further food restriction. She said, "I need a miracle, and am ready to fully commit to something that will help me keep the weight off forever."

Her first protocol was AMAZING. She sustained her very low-calorie protocol, eating less than 300 calories a day for over two months. She felt fabulous, her motivation was strong, and she didn't feel any reason to stop. She lost sixty pounds—and her type II diabetes also disappeared during that first protocol.

Ann felt healthier throughout the process than she ever remembered feeling. She slept soundly, had reduced mood swings, and increased energy. Often she'd describe the way she felt as if she was levitating. She'd cry when we met because she was so grateful. During her second phase of the first protocol, she lost another five pounds and her metabolic rate went up over five hundred calories burned per day. The protocol *was* her miracle.

However, during a six-month break before her second protocol, she'd slowly gained back twenty pounds. She was embarrassed to see me before her second protocol, but because of her newfound emotional awareness she had made a significant personal discovery. She realized she hadn't fully resolved the emotional insecurity that started when she was an adolescent. She had not yet reconciled the trauma of family molestation and a mother who was in complete denial, with her drive to feel protected through food, and to feel free from the stress of her ordeal.

As she continued through her second and third protocol, her purpose was no longer physical relief. She met with a therapist in addition with the work she did with me. She knew she had to deal with the childhood cause of her emotional insecurity. She was ready to relive the trauma in her mind, and to respond differently—this time with emotional strength. In doing so, she recognized her pain was caused from the false belief that others were responsible for her emotional well-being, which is why those she trusted could cause such lifelong emotional trauma.

Taking emotional responsibility gave her the opportunity to see her past from a new perspective. She was in charge of her own life, her past no longer had the power to define her or deprive her of happiness. For years, she'd been doing what her abusers had modeled, but instead of hurting others, she had abused herself. Eating was her emotional defense, and now that she was holding herself accountable, she had no need for emotional eating.

She had already let go of alcohol, drugs and smoking, and now she was ready to let go of food. The power she previously gave her abusers, and her addictions, were now hers alone. Throughout all three of her protocols, she never cheated once, and afterwards, she focused intensely on developing emotional self-esteem, which instantly minimized her need for food. It's been over a year since she finished her last protocol, and she continues to only eat when hungry and avoids eating until full. Even as she continues to work in her bakery, she has easily maintained her weight at 124 pounds. But more importantly, she feels emotionally more secure and happier with life than she ever thought possible.

Ann's story is a case of total transformation, inside and out.

The protocol may sound too good to be true, but when you take on the responsibility for change, miracles are possible.

SECTION 2

The HCG Protocol—A Challenge of Emotional Strength

Chapter 8

Why the HCG Protocol Is Getting
So Much Attention

"No hunger! No exercise! Lose 30 pounds of belly fat in 30 days!" When most people hear claims about this type of dramatic weight loss, it's easy to understand why they'd be skeptical. It's almost embarrassing when you observe how the protocol is being prostituted by opportunistic consumers, and by the businesses that sell related products. Not only is there fear from skeptics who've never observed, witnessed, or done the protocol, but it looks even worse as proponents of the protocol debate its merits within the hCG community.

What we have is a consumer-driven medical hormonal therapy that has been applied to our superficially motivated diet mentality. People with very little knowledge in human physiology are running the hCG industry, and have no clue as to how the hCG influences the endocrine system.

Yet, even without this pertinent knowledge, the majority of people who do as Dr. Simeons directed in *Pounds & Inches,* experience results that seem to be too good to be true. Yes, they have substantial fat loss in a short period of time, fat lost in the abdominal and visceral regions of the body, lack of hunger and overall improved energy, and ease in maintaining fat loss, etc. The protocol obviously works when followed, but what makes it different than other starvation diets? With less than 500 calories, wouldn't the majority of people attempting it experience symptoms of starvation?

Based on what the thousands who've done the hCG protocol profess, the majority DO NOT experience symptoms of starvation.

How could this be true? And if it is true, shouldn't the entire world be informed? The problem is that the consumer has demanded this protocol before any reliable and relevant explanation is given. Normally scientific laboratory studies are funded by a business that pays to prove the product's effectiveness, safety, and relevance. They would then patent the product, pay representatives to educate the medical community which would, in turn, prescribe and market the product.

In this case, hCG is a human hormone that is not patentable. Pregnant women donate their urine for hCG to be extracted and used for many different hormonal therapies. So businesses that make their money creating drugs have nothing to gain. And to make matters more complex, if hCG did what almost all participants are claiming, it would reduce the consumers' need for diet pills, diets, and medications prescribed for symptoms associated with obesity. So it's up to university scientists to apply for grants in order to fund the laboratory research necessary to explain how the protocol works.

Keeping in mind the time it would take for the research, and then to publish the findings in a medical scientific journal, the consumer would have to stand by and wait—something consumers don't do well. We want instant gratification, right?

With the increasing demand from patients, and the desperate need to reduce obesity and resulting disease, many practitioners are prescribing hCG for the protocol, even when they don't fully understand how it works. But wouldn't you—considering the number of people who are obese, and have physical ailments that only fat loss can fix? Why continue to prescribe drugs that only medicate the problem when you can prescribe something that can remove the problem altogether? The demand for the hCG protocol has surpassed the scientific explanation, and this too is cause for alarm. Both Internet sales of hCG, and homeopathic hCG sales are climbing at an alarming rate.

More and more people are bypassing the medical community and choosing to be guided by businesses run on the Internet, most of which are selling hCG product without concern for the participant's outcome. The debate has grown, but without relevant argument for, or against, how the hCG works to prevent symptoms of starvation during the 500-calorie protocol. We know one thing: people who do the protocol the way it was intended, profess results that sound hard to believe. They all agree: It works, it works fast, and creates long-term results.

But because the protocol has not yet been fully approved as a safe and effective method to reduce fat, many continue to argue against it from a place of fear—and ignorance. Yet, they're okay with drastic food restrictions and medically prescribed stimulants that suppress the appetite and cause hyperactivity. They're willing to encourage the risks of surgery that force food restrictions less than the hCG protocol prescribes. Even if the hCG didn't do anything at all, wouldn't you rather people restrict their food consumption by choice, and without the risks of surgery, or heart attack-causing stimulants?

The problem with the hCG industry is that it's consumer-controlled, and without standardization it puts the consumer at risk.

It isn't the hCG protocol that's harmful, but the lack of integrity within the hCG community as it concerns the safety of the participant, and the effectiveness of their product. Because medical hCG requires a prescription, people have turned to homeopathic hCG as a more convenient and less-costly alternative. The problem: Anybody can sell homeopathic hCG, and it's not regulated by any standard.

Most people who buy homeopathic hCG on the Internet are not aware they might actually be buying plain water. The many unsubstantiated claims prey on the fact that participants assume medical research has been done to approve the product being sold, and what's been claimed. Skepticism is a good thing, particularly when it concerns Internet-purchased hCG, and any help you think you are getting from online "experts."

I always laugh when I hear someone call themselves an expert of the hCG protocol, especially when I hear them explain how the protocol works, using Dr. Simeons' theories. They may be an expert but it's "an expert" at reading and memorizing the protocol, not necessarily understanding the physiological process of the protocol, and how it could influence the entire endocrine system. In fact, every single person who's ever done the protocol correctly is an expert. That isn't saying much because it's not hard to understand eating less than 500 calories in food.

The numbers of blogs, websites, and Internet support sites are over-whelming, and many of them conflict with each other regarding the information they provide. Some say you can mix your vegetables, others claim you can drink artificial sweeteners. Now we have new methods injecting hCG only once or twice a week, and others that have added calories to the protocol to assuage the fear of skeptics.

But there's more good press than bad, which makes it an easier sell. The people who are the most fearful may have money and careers to lose because of the success and popularity of the hCG protocol. Rumors have been started that make absolutely no sense, such as claiming hCG makes a man grow breasts. This is a hilarious, considering most men take hCG so their testicles will work while they're taking steroids, or to make un-descended testicles mature. HCG is not a sex hormone— this claim obviously makes no sense. But until some legitimate proof exists, expect false claims, false advertising, and scams.

Without proper scientific laboratory investigation and explanations that makes physiological sense, there are no "experts," and there is no "guarantee" it works the way we're told it does. It's all speculation, including my attempt at a more current scientific explanation in this book. We're all making assumptions, and basing everything solely on observation. Let's hope by starting this conversation, we can at least speed up the process and motivate research, sooner rather than later.

Chapter 9

How Something So Easy Could Be So Hard

Don't expect the protocol to be easy, because it's not.

You could be hungry if you don't follow the protocol properly, and you're going to struggle with emotional hardships, if not during the protocol, then after. Those who've done the protocol know what I'm talking about.

- How many of you actually know you did the protocol correctly?

- Did you do the "gorging days" while you were taking the hCG—or mistakenly before taking the hCG? (Gorging should be done *with* hCG administration.)

- Did you use hCG that was legitimate human Chorionic Gonadotropin, and where did it come from?

- Did you use the appropriate amount, and do you understand the importance of not getting too much?

- Did you obsessively weigh yourself? And did that help keep you motivated?

- Did you cheat?

Many factors must fall into place for the protocol to be done properly, and most who've done the protocol successfully, know weight is lost easily. However, the choice to follow the protocol properly when one has a challenge is difficult because not all protocols are created equal.

Here's a question for hCG protocol veterans and businesses: Have you noticed people who've successfully completed their first protocol tend to struggle and cheat more often during subsequent protocols?

They profess the protocol is the most amazing weight-loss process, and refuse to do anything different. After experiencing the miracle of the hCG protocol, it just doesn't make sense to attempt anything else. *Nothing compares!* Yet, once they've followed the protocol successfully, and know how easy it is for the body to reduce fat, following the protocol again can be more difficult.

For many, this is because weight loss is very easy. They think they can effortlessly recover from cheating. But this justification shows emotional insecurity, and more important, it shows little regard for the hormonal implications to the entire body.

People justify cheating on the protocol all the time, not because of hunger, but because they want to continue to eat without limit like the rest of our culture. They convince themselves cheating is okay because they can lose the resulting fat during the following low-calorie days. But after even just one "cheat" a week, people find they've only lost 3-6 pounds in a month's time, when they easily lost well over 15-25 pounds before.

If you observe the Internet culture and description of participant's protocols, it's not uncommon to read how some people have attempted the protocol many times—some over ten times. And that's not because they have 300 pounds to lose, but because they can't seem to follow the protocol with the same focused restraint they did the first time.

I believe part of the problem is the over-emphasis on lack of hunger as a selling point. If you think about it, most of us who have too much fat eat without hunger all the time. Why would lack of hunger during the protocol make any difference? It really doesn't. Participants who do the protocol correctly complain very little about "real" hunger. But when it comes to emotional hunger and the desire to eat, they complain more, especially if they don't lose weight as fast as they'd like. If they aren't rewarded with weight loss to make up for their emotional restraint, they are more likely to cheat.

What's interesting is that fat loss during the protocol is significantly faster than other diets (3-8 lbs. of fat loss a week is normal) and still, it's not fast enough to compensate for the more intense emotional hardship, when compared to other weight loss programs. The biggest vulnerabilities with this protocol are the participants' emotional need for food, and their unwillingness to feel emotional distress, without needing to cheat for security.

Even with all of physical benefits, deprivation from eating emotionally can cause irritation, anger, sadness, and feelings of emotional weakness and vulnerability. This is why monitoring weight becomes an obsession for most during the very low-calorie protocol. The immediate drop in weight gives a lot of reward, and many feel an emotional high. The protocol actually works—compared to all the failures they've experienced. But as soon as there's a day when the scale doesn't move, sadness, confusion, despondence, and anger can result. All of a sudden, what was a massive high becomes a catastrophic low, and eating less feels like a burden.

For many people, one cheat marks the end of their protocol. And if they continue to cheat day after day, they wonder why they have hot flashes, night sweats, sleep disturbance, and more anxiety, and why they feel tired, depressed, and discouraged, as they gain weight while eating fewer than 1000 calories.

I've even had participants blame the protocol for their fat gain because other diets that don't cause such sensitivity to gain when they cheat. The protocol is *not a diet*! Diets can be cheated on without the immediate gain in weight. But the problem is not the protocol, but rather the participant lack of understanding or half-hearted effort.

It's as if we're dealing with entitled adolescents who eat anything they want, whenever they want, but complain about the consequences, hate the fat, and blame their body for malfunctioning.

I encourage each participant to assess his or her desire to eat emotionally before attempting the protocol. Could you eat less than 500 calories without feeling emotionally deprived and needing to weigh yourself for motivation? I suggest you attempt to eat only with hunger for a couple of weeks before you decide to start the hCG protocol. This may help you recognize where you may struggle emotionally before you start the process.

- What are you going to do at night if you're not allowed to eat for entertainment or to decompress from work?

- How will it feel when you are surrounded by people who are liberally eating foods that you can smell, and normally would indulge?

- How are you going to handle travel, dates, meetings, and environments where eating is customary?

- Are you willing to give up alcohol for weeks on end?

And more importantly, do you really understand how the protocol works? Are you aware that by giving yourself the correct amount of hCG that you are directly stimulating the hormones that will make you extremely sensitive to fat gain and subsequent hormonal problems if you cheat?

The hCG protocol is a hormonal therapy, and it needs to be followed with strict adherence. Thoroughly prepare yourself before you start, and know exactly what you're starting.

Chapter 10

The Emotional Challenge

"The diencephalon is an extremely robust organ in spite of its unbelievable intricacy. From an evolutionary point of view it is one of the oldest organs in our body and it's an extremely robust organ in spite of its unbelievable intricacy...This has tendered it extraordinarily adaptable to all natural exigencies, and that is one of the main reasons why the human species was able to evolve. What its evolution did not prepare it for were the conditions to which human culture and civilization now expose it."

- Dr. A.T.W. Simeons, *Pounds & Inches*

People who obsess over their weight are the biggest challenge. Not because they are aware of the excess leptin that their fat creates, which magnifies fat's fueling function, but because the foundation of their motivation is completely based on whether or not the scale says what they want or feel entitle to.

During the protocol they may sleep better, have improved energy, and increased libido. It may also eliminate hot flashes, heartburn, breathing issues, type II diabetes, etc. They may even notice their joints are less inflamed, their body temperature feels less hot, their overall mood has improved, and their clothes are now a few sizes too big. But if weight defines their motivation, when the scale doesn't reflect what they want, adhering to the protocol may not be worth further effort. People who weigh obsessively, day to day, have a higher probability of cheating. For this reason, I suggest people *should not weigh themselves* as often as Dr. Simeons prescribed.

Weight is the worst indication of anything. The most volatile fluctuation in the body is fluid and waste, which can go up and down five pounds in one day. A more accurate assessment would be body-fat composition. This measures how many pounds of fat is part of the total weight. For example, a person who weighs 200 pounds, and is 40% body fat, has eighty pounds of fat. The rest of the weight comes from his lean body, which includes bone, water, tissue, blood, waste, food you just ate, and everything else but fat. What makes a person's weight go up and down isn't 100% fat. Most fluctuations are caused by lean body weight. To put all your motivation on weight, is setting you up for an emotional roller coaster ride. If you happen to be constipated one day, and your weight doesn't go down like you thought, you may have less motivation for following the protocol.

Unlike most diets, the hCG protocol *does not work* if not done correctly. By relying on continual weight loss to motivate your adherence to the protocol, you're likely to feel deprived without it, increasing the odds you'll justify cheating. Cheating would cause immediate weight gain, negates the physiological healing benefits, and perpetuates and exacerbates an emotional-eating dependency.

Before attempting the protocol, some assessment of one's drive to eat emotionally, as well as his or her emotional drive to restrict, must be made. Unless the participant finds emotional eating a frustrating nuisance, or can find reason to eat less without having to monitor his or her weight for motivation, he or she will most likely cheat or count down the days to the end of the first phase of the protocol to indulge a long-awaited emotional reward with food.

Because the protocol is simple and direct, there is little wiggle room for cheating. And because you have longer periods of time without hunger, the justifications to cheat are magnified and more obvious to an observer. This makes it apparent when and why participants use emotions to justify eating, despite being forewarned of the physiological result—weight gain. And to make matters even more clear, when weight or size is not monitored, feelings of emotional restriction are also magnified. This platform, used to identify and resolve emotional dependence on food and develop emotional autonomy around eating, could be considered a meaningful approach to an essential "emotional eating rehab."

I've found the limiting factor to the protocol isn't the protocol itself, but is the participant's wavering motivation to adhere to the constraints required to complete it successfully. If weight is lost rapidly, following the protocol is easy. If the participant doesn't perceive his weight loss is fast enough, motivation is lessened, and the restriction feels like punishment. The protocol is appealing to those who only restrict for weight loss, because the results are dramatic, and quickly realized. The common denominator here is the necessity for the participants to monitor weight to determine adherence, rather than adhering to the protocol no matter what the result.

Those who've done multiple protocols, and cheated most of the time, can attest to the frustration, not with the protocol, but with their addiction to, and emotional dependence on food. And with the obsessive monitoring of weight, predictable emotional results can occur.

Those who struggle with their emotional dependency on eating often underestimate the severe emotional roller coaster ride experienced during the protocol, when subjects obsessively monitor weight. I've witnessed participants develop binge eating disorders and bulimia, gaining huge amounts of weight after the protocol, just to try it all over again, perpetuating a vicious cycle.

A common mistake that participants make is to assume that the protocol restrictions are appropriate without the hCG, and the same food choices will suffice indefinitely. Participants are so amazed by the first phase results that they're determined to live forever with the safety net of the protocol. With their fear of weight gain, their resolve tightens and their emotions restrict them further. Within days after their last injection, hunger returns, and their urge to eat can overtake their well-intended, but naive approach.

Many people test the second phase, cheating here and there, and watching the scale for a gain of more than two pounds. Their scapegoat is Simeons' recommended day of skipping meals, or "steak day," Skipping meals is recommended when a gain of more than two pounds is observed during the second phase. The participant skips both breakfast and lunch, and dinner is a "huge steak with only an apple or a raw tomato." Many use this recalculation as an excuse to over-consume, setting up a new emotional rollercoaster between eating too much, and continually attempting to offset the damages with a "steak day."

After the second phase is over, when sugar and starch foods are allowed more freely, even more excuses are used to justify over-consumption. Without the previously enjoyed weight loss, the motivation to eat less is hard to find. Without rules, participants find weighing themselves meaningless, and even punishing, especially if they know full well they're eating excessively and regaining fat weight.

Participants who find *emotional eating* gratifying will test their eating freedom, without any value given to their body's hormonal system and the physical consequences. Not only do they experience physical harm, but also immediate and aggressive fat gain. Many justify this behavior, knowing that within weeks they can go back to another protocol to lose (again) whatever weight they've gained.

Attempting the protocol without paying attention to reasons for emotional over-eating, and finding ways to prevent it, can perpetuate and aggravate the cycle of extreme constraint and obsessive weighing—to the unrestricted freedom of gluttony and disregard for the body. For this reason, many find that following subsequent protocols becomes more difficult, even though the parameters are exactly the same as the first.

I observed those who participated in second protocols, and no matter how well the participants maintained their losses, they found the second protocol more emotionally difficult. This was surprising, considering experience from the first protocol should improve one's grasp of the restraints. But adherence to the protocol has less to do with understanding the restraints, and more to do with a desire to comply.

Satisfaction gained from the weight lost during the first protocol inadvertently weakens the resolve and desire to restrict again. Even when there are significant amounts of fat still to be lost, previous results and newfound enjoyment in food, makes additional protocols feel even more restrictive and punishing than previous attempts, despite the lack of hunger that adherence to the protocol provides.

To conclude, no matter how miraculous the human body is, or how effective the signal of leptin is to reduce hunger, the emotional drive to eat surpasses all physical cues and ignores the physical signal not to eat. In order to make responsible emotional change, we must use a different approach to Dr. Simeons' protocol.

Finding a motivation to change that is stronger than the desire to lose weight, requires a more meaningful purpose; one that intentionally targets the source of the emotional problem. Success must be the goal, and success must be measured by a diminished emotional desire to eat, not by weight loss. For this approach, we suggest taking weight and size out of the equation, removing the scale altogether.

Combining a team of assistants that move the focus away from looks, fat, and weight, towards emotional autonomy, self-confidence, and health, would have a big impact on participant responsibility. A practitioner, dietitian, counselor, and coach teamed together with the same approach, each in his or her own areas of expertise, would dynamically and effectively assist participants in their quest for independence from emotional eating.

Participants should sincerely assess their level of behavioral addiction to emotional eating before starting the protocol. The most important question they should ask is if they are willing to accept and allow the inevitable feelings of doubt and vulnerability that arise when addictions are left behind, and emotional independence is tested.

The weight-loss industry doesn't need another fad diet. It needs a revolutionary approach that not only effectively rehabilitates the metabolic compensations that predispose sensitivity to fat gain, but also addresses emotional eating disorders. This revolution could be accomplished with Dr. Simeons' hCG protocol, if both the science of human physiology and psychology find a new approach that combines to create a meaningful application.

> *"The problems of obesity are perhaps not so dramatic as the problems of cancer, or polio, but they often cause lifelong suffering. How many promising careers have been ruined by excessive fat; how many lives have been shortened. If some way—however cumbersome—can be found to cope effectively with this universal problem of modern civilized man, our world will be a happier place for countless fellow men and women."*
>
> – Dr. A.T.W. Simeons, *Pounds & Inches.*

Chapter II

The Mind:Body HCG Method: A Different Approach to the HCG Protocol

"Reality check: you can never, ever, use weight loss to solve problems that are not related to your weight. At your goal weight or not, you still have to live with yourself and deal with your problems. You will still have the same husband, the same job, the same kids, and the same life. Losing weight is not a cure for life."

– Phillip C. McGraw, *The Ultimate Weight Solution*

I love the hCG protocol. It provides the most amazing platform to experience famine without having the risk of physical starvation. You're limited to less than 500 calories a day, which would test any person's dysfunctional eating habits. But because the amount of hCG that is used powerfully stimulates leptin, hunger is minimal, and body fat sufficiently supplies fuel to prevent large dips in blood glucose.

Essentially, the protocol not only heals hormonal problems caused by excessive amounts of fat cells, but it can also be used as a platform to create a new relationship with food that prevents problems caused by dysfunctional eating.

The protocol constraints provide black and white boundaries that make it easy to direct effort towards reconnecting with true hunger. This rigidity works only because it takes such a small amount of food to relieve fleeting hunger, creating a simplified environment that makes emotional hunger obvious.

What an incredible environment to relearn how the body communicates its need for more leptin through hunger, and to learn a functional way to eat without having to actually go through a real famine.

Even so, if gluttony and dysfunction continue as the primary model we use to eat, no diet will work to reduce obesity or maintain any amount of fat loss.

Our culture has shaped a social norm that supports eating large quantities of food any time, any place, and for any reason—as an entitled freedom. We have a national eating disorder. Validated gluttony has become a cultural catastrophe where diseases associated with obesity claim more deaths than at any other time.

Because we have social ideals for the way we look, obesity is targeted as the problem, NOT the social dysfunction or gluttonous actions. The weight loss industry has taken full advantage of these standards, using weight as a marketing tool and their only indication of success. They completely bypass our disordered way of thinking that creates emotionally induced consumption. Instead, all focus is on their rules and definitions of restriction, using fat loss as bait.

Those who lose weight, gain it back after returning to the cultural ideals of dysfunction and gluttony, dejected and perplexed as to why there's a "battle" with weight. We blame everything but our justified eating, believing something must be wrong with our bodies—without questioning the values we've modeled or the "weight-loss" concept.

Lack of responsibility has created a dependency on huge businesses— spending billions to feed us, and billions to remove our fat. Both enable us to remain blind to our perspective about eating. And, that is the problem; a change of mind is the only real solution.

We need to start searching deeper than the food and start asking why. If you're given more than you need, why do you eat until there's discomfort? When did we decide validated gluttony was ok, but not the resulting fat? Why do we fail to acknowledge the scale as we are gaining, but obsess over it when dieting? We have a bigger crisis, and it has nothing to do with fat and everything to do with emotional dependence.

Despite the potential to lose fat and to regain physical health, many participants cheat and find the protocol burdensome, boring, and oppressive when compared to our gluttonous culture. Many fail miserably, not because it doesn't work, but because there's something else wrong. Like the rest of the weight loss industry, there needs to be a change in how the "battle with weight" is viewed.

The reason why I created the Mind:Body HCG Method was to help expose participants to a functional approach to eating and a revolutionary way to use Dr. Simeon's protocol for that cause. Not as another diet, but as a tool to help redefine emotional self-esteem not reliant on eating.

The decision to do the protocol can be simultaneously an exciting and frightening choice. There's anxiety over whether or not it actually works, whether hunger will be a problem, or if it will ruin the metabolism even further. There's the fear of needles, hormones, and long-term side effects, as well as the skepticism as to whether or not the weight loss will maintain itself. Some people panic just thinking about giving up food, and others doubt their ability to handle life without having eating as their coping mechanism. But once the decision to do the protocol has been made, the excitement makes waiting unbearable.

However, when the choice is made to participate in the hCG protocol using the Mind:Body HCG Method, there is hesitation. There's doubt and concern about one's emotional ability. There's apprehension about feeling the vulnerability that's an inevitable part of the process. But there's also an excitement that is steadfast, calm, and has a deeper desire to find emotional independence from food.

The Mind:Body HCG Method:

- **Takes advantage of the hunger-free environment** and the limited choice of food during the protocol to focus on detaching emotionally from eating. The goal is to gain emotional autonomy, thus eliminating the justification to feed without a physical need.

- **Is an organized strategy used to reestablish a keen sense of hunger to identify and expose emotionally justified eating.** When the physical sense we all have and are born with is relearned, eating when the body has no hunger becomes an obvious behavioral problem causing detriment to the body. With the method, eating is detached from emotions stemming from weight gain or loss, dissolving the concept of "diet."

- **Removes the scale, and all the cultural baggage that goes with it.** This forces the participant to find motivation that isn't reliant on the superficial reward of weight. Instead, motivation to eat less is reinforced by the physical relief from ailments that were caused by dysfunctional eating, by liberating themselves from guilt and shame associated to eating certain foods, and by gaining a sense of emotional balance that doesn't require rigid control.

- **Measures success in how well the physical sense of hunger can be recognized** and used as a guide. When hunger is differentiated, it can be used as a tool to observe dysfunctional eating—justified and modeled by family, culture, and diets. A keen sense of hunger is the key for when, and how much, to eat after the protocol is over. Functional application of this sense, when huge quantities of uncontrolled food options are available, is the ultimate physical and emotional success. Eating less, not for the reward of weight loss, but because the body needs less.

- **Starts by identifying and observing hunger and satiety without any emotional motive**. To do this the hunger/satiation scale must be defined and used before and after each time drink or food is put into the body.

The hunger/satiation scale is a two-part scale used to measure and define the physical sense of urgency to eat, as well as the volume and symptoms of excess after hunger is gone. Here I define each number in my own words. Throughout the process, participants should use their own words and descriptions.

Remember, hunger is in terms of urgency, and fullness is in terms of physical sensation.

The hunger scale

1. *Disparaging.* Hunger is actually subsiding, as you feel less energy, less focus, and less desire for movement. Your headache continues, you feel lightheaded, and your stomach could have an acidic feeling. You're a bit cold, and your posture is lazy and rounded. You feel shaky, and a bit nauseous.

2. *Critical.* You have anger, irritability, your head hurts, and you don't care what food you eat as long as it's in large quantity—and fast (you're craving starches and sugars, combined with fat).

3. *Urgent.* You're uncomfortable, and you should have eaten ten minutes ago. Search for food is now imminent, as hunger is increasingly urgent, and choice of food is becoming less rational. Fast food is very appealing, and traditional restaurants seem less tolerable.

4. *Patient.* You're hungry, but can wait a bit. This is a good time to start prepping food for a meal. Most people can tolerate the wait in a restaurant at this point.

5. *Content.* You feel nothing, perfectly comfortable with or without food. Hunger is completely gone, with no sense of urgency.

The fullness scale

6. *Satisfaction.* You are confident your hunger is gone. You are feeling good.

7. *Satiation.* You're feeling a bit too satisfied, burping, and feeling some discomfort in the belly. Because you have no hunger, continuing to eat would mean that you've justified it emotionally. This is usually when a dieter feels guilty, and may defend a compensatory binge.

8. *Full.* You're uncomfortable, and definitely feeling your stomach. There's still some room for food because the stomach hasn't started to stretch yet. Will need to wait three or more hours until your next meal.

9. *Discomfort.* You're very full, and feeling sick. Stomach is distended, with no more room for anything. Perhaps you have indigestion and a headache, and wish to lie down to reduce discomfort.

10. *Pain.* You've eaten so much you're contemplating inducing yourself to vomit in order to relieve the physical pain. You have to unbutton the first button on your pants, and can hardly stand to move. You're tired and need to take a nap, making it feel like being "Thanksgiving full."

This scale does not determine how fast hunger goes from feeling one number to the next. How quickly hunger elevates depends on the individual, and how much fat he or she has relative to activity level. The same goes for how long a person goes without feeling hunger, and how much food a person needs to feel satisfied.

Using this scale allows participants to gage the length of time they may need to wait before prioritizing food preparation and eating. They can also better recognize when hunger is gone and the feelings of satiation set in. Those who stay between numbers 3.5–5.5 (no matter what they eat) rarely feel extreme hunger or any sense of fullness. Remember, the speed at which this signal moves is dependent on the content of the food, relative to leptin sensitivity of the body and expenditure.

For those who haven't a clue if they're hungry or not, the Mind:Body HCG Method starts with the intent to establish and hone this physical sense. Once this hunger sense is recognizable, the focus is on creating a new perspective that gives participants the confidence to process their emotions independently—without food. This approach motivates people to open their minds to vulnerability—in order to intentionally test emotional independence.

The Mind:Body HCG Method Objectives

1. To identify stages of physical satiation or complete fullness.
2. To identify stages of physical hunger.
3. To differentiate mental hunger from physical hunger.
4. To identify triggers that prompt mental hunger and emotionally validated eating.
5. To allow discomfort without justification or action.
6. To eat based on physical need, reconditioning functional eating in uncontrolled situation.

1. **Identify stages of satiation and complete fullness**: Relearning to understand physical signals and signs of satiation (removed from an emotional drive) is an integral part of the Mind:Body HCG Method. For people who are completely driven to eat based on emotions or strict adherence to rules of a diet, physical satiation is either linked

to an emotional binge or freedom from the rigidity of the diet. When forced to eat to full capacity without emotional validation, the physical detriment is much easier to identify. This objective is relearned during the loading days, at the start of the protocol.

We recommend you choose days that don't make sense at all. Refrain from loading at a birthday party or a cultural holiday feast. Try not to have a good reason to force-feed and this will help you relearn the body's signals separate from the dysfunctional drive. On the satiety scale, this would be identifying 6, 7, 8, 9, and 10. Clearly differentiating a 5 from a 6 is imperative in relearning what it means when there's enough leptin communicating when food is no longer needed hormonally and when to stop eating (no matter what the serving size is.)

2. **Identify stages of physical hunger**: Hunger for many has never been felt. In fact, it is not uncommon for people to think that hunger is when they no longer feel full. Relearning to recognize these physical signals along with the true feelings of hunger is also vital in relearning to eat and gaining independence from our culture of gluttony, and the rigid concept defined by a diet.

    Using the hunger scale, urgency to eat is identified using numbers 1, 2, 3, 4, and 5. This range of hunger goes from feelings of nothing (5) to complete loss of concentration, irritability, and painfully intense hunger (1). Clearly understanding the difference between 3 and 4 is critical in relearning the communication that leptin provides as a signal for when the body needs food hormonally, and when to start eating.

3. **Differentiate mental hunger from physical hunger**: During the protocol mental hunger becomes clear as the rules are less about weight and more about identifying habits, emotions, and dysfunctional ideals centered on eating and food. When physical hunger is clearly defined (removed of emotional drive) mental appetite and

hunger are distinguishable and brought to the forefront of consciousness. This allows the participant to choose to evaluate where the mental hunger came from, and to choose not to eat when there is no physical need.

4. **Identify triggers that prompt mental hunger and emotionally validated eating:** Some people know exactly why they eat, but they may not realize why it's emotionally rationalized. These are the triggers that create mental appetite and emotional hunger. For most, eating compensates for feelings of insecurity that is created by doubt in their ability to create emotions independently without food. Whatever the trigger, if no physical need exists, the mental hunger requires some form of justification to give in to the dependent feeling.

5. **Allow discomfort without justification or action.** For many, the emotional insecurity that brings about vulnerability feels unbearable. Letting go of an eating behavior that has felt emotionally important and rewarding can be extremely frustrating. Many cheat, telling themselves that weight isn't important. These people become discouraged with their weight but take advantage of any grey area a diet allows. However, there is no grey area on the protocol—which makes emotional eating dependence very obvious.

During the protocol participants should seek to understand why they eat for the wrong reasons, when those desires are likely to arise, how they normally justify eating without hunger, and what they tell themselves to feel better about the physical abuse they're inflicting on their bodies. They must want to end the emotional eating dependency they've grown to enjoy, find predictable and realize creates the reflection they see in the mirror. The goal is to invite vulnerable discomfort rather than to justify removing it. Instead of eating, they must learn to acknowledge their lack of hunger and allow confidence to grow without eating as a crutch.

This concept is foreign to what's been taught to most of us since childhood—using band-aids, pacifiers, and any object to change and distract ourselves from uncomfortable feelings. Removing a food pacifier requires some internal reflection and temporary discomfort, but in the end allows for a completely new perspective and a "rebooting" of the mind and body. The goal here is to surrender all justifications to find the truth. It is to intentionally test emotional independence without food.

6. **Eat based on physical need, reconditioning functional eating in any uncontrolled situation:** Once there's a new sense of emotional autonomy it's important to test functional eating without the protocol boundaries. Building confidence in one's ability to sense hunger and eat various types of food (in different environments) is vital in the process towards emotional independence.

The second phase of the protocol—when hCG is gone and the boundaries change—gives the participants the perfect opportunity to see how well they have learned the hunger scale during the very low-calorie protocol. It reveals whether they still recognize mental hunger when the boundaries are removed, and how well they are able to filter out the old dysfunctional excuses. For many, it takes multiple protocols to completely relearn and recondition functional eating. But in the end, *weight shouldn't define success nor should eating define emotional security.*

The hCG protocol is simple but not easy. You have limited hunger, the boundaries are clear and finite, there's not much to prepare, and it's easy to understand. However, the protocol is difficult to follow when mental hunger makes situations seem arduous, and reservations about the strength of your emotional security justify cheating.

For this reason it's common to hear people recommend isolation from any temptation. They suggest that those on the protocol steer clear of social environments, stay home, and avoid restaurants during the very low-calorie protocol. *Does this sound like good preparation for emotional security in our gluttonous culture after the protocol is over?*

Unfortunately, when a diet's restraint is considered as a necessary part of emotional control, participants can go through an all or nothing emotional pendulum swing when the protocol is over.

Vulnerability, without direct suggestions for what, and how much to eat, can make the transition from the controlled first phase, to the relative freedom after the second phase, feel undisciplined and overwhelming.

When there's no protocol, no boundaries, and no clear rules, participants lack confidence in their emotional strength. Avoiding temptation keeps them weak to sense emotions independently and removes the opportunity to learn to enjoy food functionally.

It's not that we shouldn't like or take pleasure in food, but that we should become skilled at eating without making dysfunctional gluttony the gratifying focus. There's deep satisfaction with the delights and taste of food—when it's functionally eaten without guilt or overcompensation for a lack of freedom, and when the body feels absolutely no symptoms of discomfort from fullness.

For this reason, we suggest participants seek out vulnerability and emotional tests of independence. During the first phase we encourage participants to test out restaurants, parties, and every social occasion possible to gain social confidence without food. We've found this strategy allows for less fear when the first phase is over and develops self-confidence without the restraints of the protocol. It gives them a chance to retrain and recondition their habits so food is not the center of their life, entertainment, emotions, or their happiness. Detrain the old and retrain the new!

The approach and purpose behind the protocol can either be to continue the weight loss battle or end it completely. If approached with the intent to end emotional eating, then there's tolerance for discomfort as old patterns of insecurity are independently secured. On the other hand, if the focus is on how much weight is lost, the protocol will be another set of rules, a daily battle with the scale, continued punishment for having to restrict, and the constant fear weight will be regained.

No matter how miraculous the physical result, it is the participant's choice when it comes to their approach to the protocol. When applied with the hCG protocol, the Mind:Body HCG Method accelerates the process for those willing to expose the roots of their emotional dependency, deeply wanting a desire for change, and intentionally seeking to find independence without emotional help.

The Method holds participants accountable for a source of motivation that is not limited by a fleeting resolve for weight reduction. It exposes dysfunctional eating and other compensations for insecurity, giving merit to authentic and genuine emotional strength. This process gives you an opportunity to break away from a life of gluttony and dieting, and freedom from the emotional limitations of both. The result is emotional strength and a body set free from the torments of abuse.

"If hunger is not the problem, then eating is not the solution."

– Author Unknown

Chapter 12

Profound Reasons to Eat Less

"We're the country that has more food to eat than any other country in the world, and with more diets to keep us from eating it."

– Author Unknown

After personally assisting participants in over 500 protocols, and witnessing thousands more, I have seen some extraordinary results. The obvious outcome when fat is lost is an improved hormonal system of the body. In short, I believe the protocol should be used to heal any hormonal problem that is linked to an over-abundance of leptin, magnified by fat, and caused, for the most part, by eating without hunger.

Here is a list of just some of the symptoms that have been linked to too much leptin:

- Polycystic Ovarian Syndrome (PCOS)
- Type II diabetes
- Heart disease
- High LDL cholesterol
- Difficulty sleeping
- Increased testosterone in women
- Increased estrogen in men
- Symptoms of adrenal fatigue
- Hypothyroidism
- Depression

- Difficulty breathing
- Heart burn
- Joint inflammation

- Fluid retention
- Night sweats
- Hot flashes

After witnessing hundreds of people while on the protocol, and listening to their personal observations of symptoms that have disappeared after their experience, I am not surprised American people are prescribed so many medications. The majority of us are now considered obese, which means the majority of us have symptoms associated with too much leptin. But, if the fat was removed, and eating became functional, then the cause of those symptoms would disappear—and so would the need for medications.

Case Study

One client in particular, stands out as it concerns the hormonal symptoms. She was diagnosed with polycystic ovarian syndrome when she was 28 years old. As described by the U.S. National Library of Medicine, polycystic ovary syndrome (PCOS) is a condition in which there is an imbalance of a woman's female sex hormones. This hormone imbalance may cause changes in the menstrual cycle or skin, and might cause small cysts in the ovaries, trouble getting pregnant and other problems. PCOS is linked to over-stimulated leptin, and how leptin affects the gonadotropins of the ovaries.

When she met with me to start the hCG protocol, she was 31 years old, and had never experienced a regular menstrual cycle. After the first three weeks on the very low-calorie protocol, she started having regular periods. She lost over fifty pounds her first round, and continued to lose weight easily during maintenance. She also continued to experience regular and predictable periods for the first time ever. She decided to follow up with her endocrinologist.

Before starting her second protocol, she was pleased to tell me that after retesting her blood work, her endocrinologist informed her she no longer had PCOS. However, I wasn't surprised. The shape of her face changed, her hair grew in thicker, her body shape completely changed as she now had an hourglass shape, rather than her previous apple shape.

This client not only lost the fat that was causing the hormonal magnification and affecting the way she looked and felt, as well as the functioning of her ovaries, but she completely changed the way she ate—the ultimate cause of her problem.

By minimizing her intake of sugar, and eating only with hunger, she felt an incredible change hormonally. Her motivation to continue to eat functionally became a new way of life for her. She lost over 100 pounds, and just recently announced she's expecting her first child. At one point, she was told pregnancy would be improbable. I have now witnessed more than five women who were "cured" of PCOS after the hCG protocol, and by following the Mind:Body HCG Method.

Every time an obese person eats without hunger, his or her entire hormonal system overreacts, and causes compensatory symptoms. This type of reaction doesn't happen to people who have much less fat because they have fewer fat cells stimulating leptin, which is why they have less hormonal compensations.

One of the most eye-opening observations I've made is why people who have more fat aren't hungry for breakfast. I was always taught that a person's chances of gaining fat are much higher if they don't eat breakfast. Now that I've discussed this with thousands of obese people, I don't believe that at all.

Breakfast may be important nutritionally, but when hunger isn't present, it is not needed hormonally. Obviously if a person has less hunger, his or her hormonal need for food is less, which makes the nutritional value of what they do eat more important. But shouldn't hunger be the deciding factor when, and how much, a person eats? Illness associated to chronic vitamin and mineral deprivation isn't the obvious concern with obesity when you compare it to the threat of chronic hormonal inflammation. The way we've interpreted the link between lack of breakfast and obesity has been backwards.

Have you ever considered the reason people who are obese don't eat breakfast is because they're not hungry? Ask yourself: are you immediately hungry when you wake up? Do you force-feed breakfast because someone said it was important, even though you are showing no reason to eat hormonally? Need for breakfast should be solely based on hunger, similar to how we'd eat if we were rationing. The body *requires nutrition that food provides, but when you are eating without hunger, you're risking the balance of the entire endocrine system,* especially when you have more cells that magnify fat's fueling response.

- Could it be that when your melatonin level (a sleeping hormone) drops in the morning, that leptin levels conversely rise?

- Does it seem reasonable that this rise in blood leptin levels be much higher for people who have more fat cells?

- Could insatiable hunger in the middle of the night, be triggered by the plunge in leptin caused by increased melatonin?

The research is available, and these questions have answers—the people who are treating obesity just need to read about it. If you're not hungry in the morning, wait to eat until you are.

Many of the symptoms that we medicate for are directly caused by eating without hunger, and this magnifies the hormonal system's response, especially when linked to obesity. Many people are turning away from medicine, and looking into hormonal therapy to relieve their symptoms, such as bio-identical hormone treatments.

Bio-identical hormone therapy is used to identify and improve under- or over -active organs, aiming to remove specific symptoms, like hot flashes, low libido, etc. However, if a patient is obese, some of his or her hormonal imbalances might be a caused when he or she eats when not hungry, which would stimulate too much leptin. For example, the meno-pausal women I've worked with notice they experience hot flashes about 15 minutes after they eat sugar, drink wine, or consume anything—when they are without hunger. By eating functionally, and minimizing powerful foods such as sugar and starch, women may notice a reduction of hot flashes. But when you combine functional eating with fat loss, many of their previous hormonal symptoms are completely eliminated too. Again, the more fat a person has, the more pronounced is her or her hormonal response after eating without hunger.

I believe bio-identical hormone therapy could be a valuable form of therapy for many. But, in order to make the most of it, education and instruction on eating with hunger is vital, and the removal of excess fat is imperative. This is where the hCG protocol would be a great start-ing point for most patients whose symptoms are linked to too much leptin.

A better approach to bio-identical therapy would be to:

1) Prescribe the hCG protocol in order to reduce the fat to a level that is ideal.

2) Teach the patient to only eat when hungry in order to accurately deduce whether or not the hormonal imbalances are caused from too much leptin, or as a result of dysfunctional eating.

3) After leptin induced inflammation caused by dysfunctional eating is eliminated, if there still exists an under- or over-active organ, then prescribe the appropriate therapy.

This approach holds the patient accountable for what she or he may be doing that causes the hormonal symptom. But even if the protocol is not used to reduce a person's body fat, just by using hunger to eat function-ally, symptoms as they relate to leptin may disappear. This observation in itself would reduce the need for pharmaceutical drugs or hormonal replacements that only remove the symptom, but don't solve the actual cause of the problem.

Unfortunately, I've found that most professionals in the medical community who want to implement a weight-loss option for their patients don't realize obesity isn't just a physical symptom to be medicated. There is a significant behavioral component that requires motivational support, and that's where most hCG programs and all diets fall short. However, I believe this is where the personal training industry would be an important part of the process.

When you add the hCG protocol to a practice, doctors are entering completely new territory: the weight loss industry. I think the key to helping patients through the protocol, and making the protocol a successful addition to a practice, is having the right support team. Unfortunately, when you leave the monitoring of your patient to a nurse, the nurse is completely out of her or his field of education and training.

Doctors should consider that it would be a great value to add degreed and certified trainers to their practice who are experienced and educated in knowing how to motivate lifestyle change. People who can also assess and understand metabolic testing and body fat composition, exercise, and nutrition and are knowledgeable in the physiology of the body's fueling systems.

Adding such a professional to your team provides huge benefits to a program, since they are able to spend the necessary time explaining, teaching, motivating, and monitoring participants. A doctor may be willing to prescribe hCG but they may not be willing to market, manage, and know how to help people who cheat, people who obsess over the scale, people who need more hand holding because they feel emotionally deprived without food, etc.

Thousands of trainers are certified, experienced, and degreed in exercise science, and want to assist clients through something as challenging as eating issues, body image, and weight loss. Because of their background implementing weight-loss programs, they already know the emotional pitfalls to expect from participants. Teaming up with a qualified personal trainer would be an invaluable service to the patient.

But before anybody can effectively teach or assist participants through the protocol we should understand how the hCG protocol works to best comprehend what it can be used for. And for this to occur, we desperately need new research.

The hCG protocol needs to be researched and applied to the new scientific understanding of leptin. Laboratory research should compare blood leptin levels, thyroid hormones, adrenal response, body fat composition, resting energy expenditure and more, during the hCG protocol. But first we require a credible hypothesis that explains why hCG can be used to alter the hormonal system of the body to avoid starvation when combined with the very low-calorie protocol.

The hCG protocol could be the ultimate therapy to revolutionize the way our culture eats, healing the symptoms of disease that are associated with excess fat, and improving our country's overall health, wellness, and quality of life. Until the hCG protocol is proven an effective therapy for the hormonal imbalances caused by obesity and dysfunctional eating, we're stuck with drugs that just mask the problem, and those drugs may cause other symptoms that require even more drugs that cause even more symptoms and round we go. We have profound physical problems caused by our desire to eat too much. Ultimately, we need compelling reasons to eat less, and compelling solutions. The hCG protocol just might provide one.

"More die in the United States of too much food than of too little."

– John Kenneth Galbraith, *The Affluent Society*

Chapter 13

Eating Like a Child

I believe the hCG protocol is like a slap in the face. It can knock some sense into the reasons why we eat, and can help us recognize the magnitude of the consequences caused by our gluttonous culture.

How many pharmaceutical drugs are being used to alleviate life-threatening symptoms caused by dysfunctional eating—and the hormonal result of obesity? How many people live day to day with symptoms that influence the way they feel, their outlook on life, and alters the choices they make that affects their future? After decades of one dieting attempt after another, it's time to stop looking for someone to spoon-feed us, and it's time to start holding ourselves accountable.

Many of my clients have expressed that they feel "out of control." They're disgusted by the way they look and feel, and are desperate for an abrupt change. Others are confused, and don't know how they acquired so much fat. They don't know which diet to follow, and can't seem to pull together enough energy or motivation to do anything about it.

Many are looking for something drastic, which is why they are attracted to the protocol. When compared to other diets, it's quite a severe alternative. They insist that the hCG protocol is their last and final effort before they throw in the towel, succumb to the body they believe is destined to be obese, and surrender to their battle with weight.

But when I describe, what for me, was a lifesaving childhood memory, they see some hope for redemption.

My Story

I am a survivor of an obsessive-compulsive disorder. I was obsessed with dieting and avoiding fat gain, which I believe was more of a body image disorder than an eating disorder. While some people are addicted to drugs or alcohol, my addiction was controlling the fat on my body.

Here's what happened:

- I would exercise for hours to prevent fat gain.

- I feared food as if it was the plague.

- My emotional disorder made my life miserable, but I was so fearful about how I thought life would be without it.

- I felt such sadness and helplessness that suicide and death seemed a realistic solution to end the psychotic suffering I self-inflicted.

In desperation, I used to fantasize what it would be like to be a child again. One memory that I would visualize stands out. I was playing outside with my brother's plastic green army figures, giving each one a name as I lined them up. Some of the figures would die, and I'd bury them; others would march around and command other figures to do things. What made this memory so attractive to me? How I felt. I was having so much fun creating with these toys that I remember vividly what it felt like to not want to come in for lunch. Mother called me in, but I just didn't want to stop playing. Eventually, when I could no longer ignore my hunger, I ran inside. Not wanting to waste another minute of my creative time eating, I didn't care what I ate, and ate very little of what Mom fixed.

I remind clients what it was like to be a child, to play for hours, and only stopping to eat when hunger couldn't be ignored. It reminds them of what it feels like to be emotionally liberated from food and dieting, when they didn't have guilt over what they ate—and how good it felt to never feel full.

For me, later on, the more I felt liberated by remembering the emotional freedom I had as a child—the more I wanted it. I decided I would eat, just like I was a child.

1)   Eat only when I was hungry,

2)   Eat whatever I wanted without judging the food as "good" or "bad," and

3)   Never eat more than I needed.

I stopped weighing myself, and decided if I were going to continue to live, I would have to unconditionally love my body, no matter what the outcome.

This emotional freedom from eating, combined with the intrinsic use and trust of hunger, saved my life, and I haven't stopped "eating like a child."

Redemption is waiting for you, and all it takes is a willing desire to let go of the idea that eating or dieting has emotional value. The protocol is your tool—a way to knock some sense back into the way you eat—to eat like a child again.

SECTION 3

The HCG Protocol—What Every Doctor, Participant, and Skeptic Needs to Know

The following chapters are meant to explain and start a new discussion about the hCG protocol—why it works, how a small dose of hCG prevents hormonal symptoms of starvation, and how to do the protocol as Dr. Simeons prescribed.

Today most of the information people get about the protocol is from the Internet. Because no other explanation exists other than what Dr. Simeons wrote in *Pounds & Inches*, a large amount of speculation, skepticism, and inaccurate assumptions (based on Simeons' theories and observations) have developed for why and how it works—or doesn't work.

But because *Pounds & Inches* was written in 1967, Dr. Simeons could NOT have formulated a theory to explain his observations in a way that makes sense, based on today's level of scientific understanding. In other words, *Pounds & Inches* is outdated.

To argue for or against Simeons' theories is a waste of time. We need to skip over how he thought it worked, and start anew. We need a new explanation for why such a small amount of hCG could prevent starvation—not applied to the invalid assumption based on calories, but applied to the modern and proven science that describes starvation as a consequence regulated by hormones.

Some of the following chapters were specifically written for doctors or practitioners, which make it very complicated for the average college graduate. I do attempt to describe why it works in a user-friendly way, but in order for any of it to makes sense, you have to let go of the outdated idea *that starvation is based on calories.*

This conversation will force the entire weight-loss and diet industry to change. Not only because the more people who do the protocol correctly will significantly reduce the occurrence of obesity, but because new science destroys the outdated idea of "calories in, calories out." No more counting, no more measuring. This is a weight-loss apocalypse.

Chapter 14

Why Dr. Simeons Thought It Worked

Dr. A.T.W. Simeons observed hCG for the treatment of obesity for decades, publishing his findings during the 1950s. In 1967, Dr. Simeons privately published his book, *Pounds & Inches: A New Approach to Obesity*, after pressing interest and curiosity from both physicians and participants. This book described his history with hCG, his observations, his theories, and his protocol.

To better understand the protocol, we recommend that you read *Pounds & Inches,* found in PDF format on our website *www.mindbodyhcg.com*.

Here is a very brief explanation of Simeons' protocol and theory.

Human Chorionic Gonadotropin (hCG) is a placental peptide hormone produced and released during pregnancy. Dr. Simeons theorized that this hormone alters the hypothalamus in the brain, directing the body to burn fat effectively during pregnancy, ensuring a continual feed of energy to both mother and baby. He believed that the body, both male and female, could re-create this fat-burning cycle outside of pregnancy, by introducing injections of the hCG hormone. His theory was that using hCG would allow obese participants to severely reduce their food intake, forcing the body to use stored fat as energy sources. He believed

125-200 units of hCG would prevent hunger, and symptoms of starvation. He implemented his protocol, and observed participants lost an average of one pound each day.

His protocol is broken down into two phases. During the first phase that is three to six weeks in duration, hCG is administered once a day through injection, and food is rigidly controlled by the very low-calorie protocol (VLCP). The second phase follows for three weeks in duration after the first, but is administered without hCG, and uses a different set of food constraints.

During the first days of injections, Simeons required his participants to complete a "forced feeding" of high-fat foods. Some people call this their "gorging" or "loading" days. After loading, the VLCP abruptly starts on the forth injection day.

The VLCP, described in more detail in *Pounds & Inches*, allows for controlled servings of low fat (if not fat-free) protein, fruit, vegetables, and crackers that total less than 500 calories per day. Over decades of observation, and through trial and error, Dr. Simeons determined the optimal specificity of the VLCP to be followed, even down to the grams of meat to be consumed. Simeons believed that as hCG allowed fat to be readily available as energy, hunger would subside, and participants could continue the VLCP for three to six weeks. Towards the end of the cycle, as fat is significantly reduced, his theory was that hunger would resume, signaling the body's acclimation to hCG. This would signify the end of the first phase, and the transition into the second phase.

Once the hCG is no longer active in the body, the second phase begins. This phase is more liberal, with increased portions, and food choices constrained to minimize sugar and starch. He purposefully asked participants to eat more to maintain their weight, rather than intentionally restrict, to continue weight loss. Participants are instructed to weigh themselves daily, and to avoid weight gains of more than two pounds. Many people call this the "maintenance phase."

I don't necessarily advocate that you follow Simeons' suggestion to carry a scale with you everywhere you go or to even weigh yourself every day. By obsessing over the scale, people set themselves up for problems as they are easily discouraged when the scale doesn't move, or fluctuates with lean body mass changes (fluid shifting). Getting a body fat composition evaluation once a month would be of greater value.

Dr. Simeons believed that during the second phase, the hypothalamus (the brain's hormonal regulatory center) would "re-set" itself, and a new, higher metabolic rate would be the result. He recommended waiting a minimum of three weeks after the second phase before starting another round of the protocol. As participants do more rounds of the protocol, the wait time between protocols should increase.

Simeons didn't explain some peculiarities in his protocol, such as why he:

- Recommended that participants choose only one vegetable to be eaten each serving,

- Suggested eliminating oil-based make-up, soaps, and lotions,

- Always skipped injections for female participants during menstruation,

- Didn't give an explanation for why an "apple-day" or second phase "steak-day" works.

Again, for full details of Dr. Simeons' protocol refer to *Pounds & Inches.*

Dr. Simeons' Theories

Dr. Simeons theorized we have three types of body fat: structural fat, normal fat, and abnormal fat, described here:

- *Structural Fat:* Cushions and supports the body and organs, keeps skin smooth and taut, and supports many other important body functions.

- *Normal Fat Reserve:* Immediate fuel that the body can freely draw upon when food intake is insufficient to meet demand.

- *Abnormal Fat Reserve:* A potential reserve of fuel; unlike the normal reserves, it is not available to the body in a nutritional emergency. It is locked away, so to speak, to be released only during extreme starvation, or pregnancy.

Simeons observed that regular dieting pulls energy from normal and structural fat reserves, as well as lean tissue. Fat loss would first be noticeable in the joints, face, and neck, leaving abnormal areas of less used fat. He used the example of bank accounts to explain the difference between normal and abnormal fat reserves. Normal fat reserves are like a checking account. Calories can come in and out any time of the day as needed, like a debit or credit in your checking account. He compared abnormal fat reserves to a savings account, easy to deposit into, but more difficult to withdraw.

Dr. Simeons believed abnormal areas of fat are almost impossible to retrieve for energy during regular caloric restriction or exercise. This fat usually accumulates in the belly, ribs, thighs, hips, ankles, and even the back of the knees, the lower back, or the back of the neck.

Dr. Simeons theorized specifically that obesity was a disease that resulted from an off-kilter regulation of the body's energy system from a malfunction inside of the brain—specifically within the hypothalamus. He disputed that the cause of the dysfunction did not stem from the sex glands, the thyroid gland, pituitary gland, adrenals, or from activity levels or caloric consumption.

Dr. Simeons believed this dysfunction of the brain's hypothalamus had three basic causes:

1. Passed down through genetics, the hypothalamus pre-sets energy balance with less food, making one's food consumption less necessary, no matter what his or her caloric demand. If the hypothalamus sets the body's energy balance to require less food, a person will have a propensity for fat gain, even when eating only what is calorically required, or less.

2. The hypothalamus can be manipulated out of a normal energy balance rate by compensating for over- or under-active organs. He states in *Pounds & Inches*:

   "It seems to be a general rule that when one of the many diencephalic centers is particularly overtaxed, it tries to increase its capacity at the expense of other centers."

For example, as the ovaries lose function through menopause, lack of ovarian hormones (gonadotropins) taxes this center of the brain to create more gonadotropin-releasing hormones. Without response from the ovaries, the body requires compensations to maintain balance. One of these compensations could result in fat gain. Dr. Simeons believed the insurgence of fat in this way happens regardless of dietary restriction. In fact, he believed enforced dieting would deplete essential and normal fat stores, and would be a disadvantage to the patient's general health.

3. With sudden bouts of stress to this system, either through rapid food consumption; dramatic sedentary or active alterations to daily expenditure; severe food restriction; or change from a cold to warm living environments, the hypothalamus is forced to compensate. These compensations are more permanent with chronic adversity. Illness, starvation, or chronic over-taxation would require the hypothalamus to adjust its set response to energy demand, requiring less food, or, on the other hand, more food.

He gave example after example of his observations, stating that weight alone wasn't a satisfactory criterion to determine whether a person was suffering from this hypothalamic disorder. One of the criteria he suggested was for doctors to observe the location of body fat. For example, with patients he thought had this disorder, he observed the "Duchess' Hump," as a little pad just below the nape of the neck; abdominal fat; breast fat in both male and female patients; and fat on the knees, hips, thighs, upper arms, chin and shoulders. These areas of fat might appear in patients who have statistically normal weight. He observed people whose body shape was irregular, like someone who is very thin in her upper body, but has excessive fat in her lower body. He suggested they too might have this obesity disorder stemming from the hypothalamus.

On the other hand, he observed people who were indeed overweight, but their fat was not distributed in areas typical of those with the hypothalamic disorder. These patients lost weight with increased exercise or typical reductions in food.

Early during his studies, Dr. Simeons thought the problem of obesity stemmed from the anterior pituitary gland, but changed his hypothesis to the hypothalamus after he observed patients who were given daily doses of hCG. Observing thousands of patients, Simeons believed hCG influenced the hypothalamus in a way that would redirect fuel usage towards abnormal fat, and with his protocol, could correct this hypothalamic dysfunction. With adequate time—a minimum of twenty-three injection days, his hCG protocol would reset a new, more appropriate energy balance, allowing for normal food consumption without gain.

Through deduction, he created the specificity ascribed to the first and second phase of the protocol. During the first phase, he believed that as abnormal fat cells become available, and as sufficient energy was supplied to the body, lean tissue loss would be prevented. Thus, his protocol would help to maintain a healthy muscle mass and metabolism during the very low-calorie protocol.

Dr. Simeons also theorized that with the unlocked potential of abnormal fat, severe restriction hunger diminishes as fat energy sufficiently "feeds" the body. Essentially, as fat sustains energy, the sensation of hunger would weaken and be experienced less often.

He believed that during the second phase of the protocol, the participant's hypothalamus would reset the body's fat-burning rate to function faster, making the second phase vitally important as therapy for the brain's hypothalamic disorder.

Considering the level of scientific knowledge available when he wrote his manuscript, Simeons' detailed observations and his theories are quite amazing. By today's standards, however, they're insufficient.

Chapter 15

For Doctors and Practitioners:
A New Hypothesis

*"I have never had an opportunity of conducting the labo-
ratory investigations which are so necessary for a theo-
retical understanding of clinical observations, and I can
only hope that those more fortunately placed will in time
be able to fill this gap."*

– Dr. A.T.W. Simeons, *Pounds & Inches*

HCG Protocol: A New Hypothesis

HCG injections of between 100IU and 150IU might stimulate, from
fat cells, sufficient amounts of leptin to prevent symptoms of starvation imposed by the very low-calorie protocol (VLCP), as described
by Dr. Simeons. The controlled hCG/protocol environment should
stimulate enough leptin to minimize hunger, to maintain a normal and
healthy thyroid signal, and to prevent significant reductions in lean
body mass by effectively maintaining energy homeostasis through
fatty acid oxidation of leptin-sensitive fat stores.

The controlled demand imposed by participant adherence to the
VLCP will prevent fat gain by offsetting the susceptibility to over-

stimulate leptin. This controlled demand with optimized fat utilization should, with time and physical adaptation, up-regulate mitochondrial biogenesis. During the second phase, as energy demand on fat is significantly reduced, new smaller mitochondria might progressively increase oxidative power, measuring an increase in resting energy expenditure (calories, per pound of body weight, per day).

Using this hypothesis as the basis of future research might explain what is observed during the protocol.

The following material is meant only to start new discussion about how hCG might influence energy homeostasis when food is removed.

Energy Homeostasis

To start, let's acknowledge the inherent difficulty of understanding the body's integrated system of organs, each requiring its own nourishment and energy demands, in addition to understanding the systems of tissues dependent upon these organs. The energy needed to sustain our organs and tissues is a system that feeds and depletes. It gives and takes from one organ to the next, all while accommodating the complex influence from both physical activity and food. This balance of energy demand and energy sharing is called energy homeostasis, and maintaining homeostasis sustains these integrated systems during both feast and famine.

All systems integrate fuel and energy demands not only daily, but over a lifetime. This constant striving for homeostasis is what stimulates the feelings of hunger that prompt us to eat, and to stop eating when we're satiated. Perhaps the most critical element in achieving homeostasis is maintaining a stable blood glucose level.

We are fed from many sources other than food; some sources of fuel are fat, muscle and liver glycogen, body protein, and blood glucose. These "tissue" fuels are not stocked equally. Some have more reserves than others. Fat and body protein by far surpass the fleeting amount of energy reserves held in both glycogen and blood glucose. The total integration of these fuel systems for short and long-term metabolic homeostasis is vital to life, hourly, and over the period of our body's life.

The energy our body captures is powered not only by food, but also by our tissue reserves. However, these substrates must be converted into what can be captured before the body can use it as energy. As you eat, the food you consume is not yet in a form that can be captured as energy, so tissue reserves are readily available to meet immediate demands. But for tissue reserves to be released, key hormones that determine when and how much energy is needed must be accessed.

Leptin is one of the most important energy-controlling hormones. Since its discovery in 1994, we more fully understand leptin's key role as a fatty acid synthase (FAS) inhibitor, and most notably as an anorexigenic hormone affecting the signal of hunger, the function of the thyroid, and fat metabolism. [2 15]

Similar to insulin, leptin levels fall and rise in coordination with blood glucose, signaling to the body and brain when and how much energy reserve is available. [3] Leptin helps maintain blood glucose levels by regulating fatty acid use in skeletal muscle for energy, and preserving blood glucose for other more important organs to use. [4]

Leptin is primarily found in white and brown fat cells, but could also be produced in the mouth, placenta, ovaries, skeletal muscle, stomach, mammary cells, bone marrow, pituitary and liver. [5 6] The rise and fall of leptin levels influence hunger, thyroid stimulus, fat metabolism, and fat gain. To successfully apply the modern science of leptin's functions to Dr. Simeons' protocol, I will discuss leptin as it relates to four areas; the hypothalamus, skeletal muscle, fat, and the thyroid.

Leptin: Bridging the gap between the hCG protocol and hunger.

As a diet begins, and food is restricted, blood glucose levels fall. As blood glucose levels drop in the body and brain, leptin also depletes. [7 12] As leptin levels fall, there is a reduction in malonyl-CoA, a recognized intermediate, in the hypothalamic-signaling pathway that controls feeding behavior and energy expenditure. [15] Recent evidence suggests that food deprivation, and the associated decrease in hypothalamic malonyl-CoA, increases the expression of neuropeptide Y (NPY) and agouti-related protein (AgRP), which produces the sensation of hunger. Conversely, as blood glucose and leptin levels rise after eating, the resulting increase in malonyl-CoA reduces the expression of NPY and AgRP, producing feelings of satiety when hunger is alleviated. [7 10]

Studies have shown administration of a fatty acid synthase (FAS) inhibitor (such as leptin) to the central nervous system in obese mice, dramatically reduces feeding behavior, with the increase in hypothalamic malonyl-CoA concentrations.[13 25] These findings show that during very low-calorie diets, a stimulant of a FAS inhibitor like leptin, would raise malonyl-CoA levels, and decrease the expression of NPY and AgRP. Theoretically, this should sustain feelings of satiation for longer periods of time with less food.

Leptin: bridging the gap between the hCG protocol and fat mobilization.

Leptin is primarily expressed and secreted by fat cells. As fat mass increases during energy surplus, blood leptin increases and interacts with its receptors in the central nervous system (CNS), leading to increased malonyl-CoA expression in the hypothalamus, and decreased hunger. [22] Although there could be fat loss due to lack of hunger with a FAS inhibitor, studies have shown central administration of FAS inhibitors transmitted to the skeletal muscle from the CNS, increases fatty acid oxidation and, with time, increases resting energy expenditure. [16 23]

As FAS inhibitors increase in skeletal muscle, the result is a decrease in muscle malonyl-CoA. This outcome essentially determines whether or not fat is used for energy. [4] Muscle malonyl-CoA is a potent allosteric inhibitor of muscle carnitine palmitoyltransferase (CPT-1). CPT-1 is like a doorway on the mitochondrial membrane, opening or shutting access for fatty acids to enter and be converted into fuel for the body. When CPT-1 is deactivated by muscle malonyl-CoA, entry of fatty acids into mitochondria for β-oxidation is inhibited. [15]

Muscular malonyl-CoA formation is catalyzed with increased activity in the enzyme Acetyl-CoA carboxylase (ACC). ACC is strongly inhibited by AMP-activated protein kinase (AMPK), which is stimulated by leptin. [11] So, as leptin levels decrease, AMPK is deactivated, which activates ACC. ACC creates malonyl-CoA, which inhibits CPT-1, and thus reduces fatty acid oxidation. [13][15]

This happens as a response during starvation when blood glucose and leptin levels fall, preserving fat for longer periods of time, and forcing muscles to use other tissue substrates instead.[12] However, eating has the opposite effect.

After eating, when blood glucose and blood leptin levels increase, the activation of AMPK deactivates ACC, which decreases muscular malonyl-CoA. As muscle malonyl-CoA declines, CPT-1 activates and opens access for fat into the mitochondria, where energy can be supplied through β-oxidation. [4][11] This might explain how eating food that is not yet in a form that can be captured as energy, stimulates the use of stored fuel for immediate use.

New science has shown that this system can be successfully manipulated, not only to counteract symptoms of starvation, but to improve metabolic rates. Centrally administered FAS inhibitors during food restriction, rapidly increases the expression of skeletal muscle peroxisome proliferator-activated receptor-α (PPARα), a transcriptional activator of fatty acid oxidizing enzymes, and uncoupling protein 3 (UPC3), a putative thermogenic mitochondrial uncoupling protein. [23][15] Daily administration of FAS inhibitors over time increases the number of mitochondria in white and red skeletal muscle. This could explain why studies show increases in metabolisms tested through indirect calorimeter. [23][26][27]

This evidence shows that if there was a way to safely increase a FAS inhibitor such as leptin, as well as create energy demand with food restriction, the response over time should be to acclimate with more mitochondria, resulting in a higher caloric-burning capacity. But without a FAS inhibitor, one should expect with the same food restriction to see a slowed loss in fat, increased loss of lean tissue reserves, and a resulting decline in resting energy expenditure.

To prevent the natural decline in fat mobilization with a very low-calorie diet, there must be an alternative way to stimulate leptin to decrease muscular malonyl-CoA, This allows fatty acids to have continuous access into the mitochondria, where fat could provide substantial fuel for the body without significantly depleting blood glucose. This optimized fat utilization would prevent the need for the body to use lean tissue reserves during extreme caloric deficits and, over time, stimulate mitochondrial biogenesis, ultimately increasing the rate at which a person burns energy fuel at rest.

Leptin: Bridging the gap between the hCG protocol and fat gain.

High levels of leptin in adipose tissue, without equally sufficient expenditure, have the opposite effect. Studies show that extremely high levels of leptin, similar to those seen in the obese, increase peroxisome proliferator-activated receptor-gamma (PPAR-gamma), which is the master control switch for fat storage. [21] [22] [14]

PPAR-gamma activates a host of enzymes that promote the esterification of fatty acids to create triacylglycerides (TAG), and advances the formation of lipid droplets from these TAG. When administered to mice, high levels of leptin increased the cellular expression of PPAR-gamma by 70-80%. [14] Leptin signals to the brain that there's ample energy in storage, but also forewarns pre-adipocites to make room for more fat cells.

The more fat a person has, the more leptin his or her body produces. [21] Essentially, if you were to compare two people who have the same exact metabolic rate, but extreme variance in body fat composition, their bodies would have a different response to the same food. If they ate the same exact amount and type of food, the more obese person would have much more blood leptin stimulated, due to their larger amount of body fat.

The excess amount of leptin, without equal excess energy expenditure, can cause an imbalance in energy homeostasis, making the body more sensitive to resulting fat gain as a need to recapture and compensate for the imbalance. Leptin's stimulus of PPAR-gamma would complement insulin as a survival mechanism to make room for more fat, aiding in the preparation for more energy storage cells as an adaptation for long-term energy homeostasis. A person with less fat would have less leptin, which might better compliment their metabolic energy balancing system, thus making him or her less sensitive to fat gain—even when eating the same exact meal as a more obese counterpart. Hence, fat gain and loss is not a linear function of calories eaten and expended because fat hormones, such as leptin, greatly influence energy homeostasis, and the body's resulting compensations.

Both fat-preserving and fat-creating effects of leptin will function to conserve fat during starvation, and to form fat when food is excessive. Leptin's fat burning and storing/preserving relationships seem to follow an "inverted U" model. Leptin's fat-conserving functions are maximum with high and low levels, and its fat-burning functions are optimal in the middle.

If leptin is stimulated by an outside influence, there might be less necessity for food and more sensitivity to over-stimulate leptin production. This excessive stimulus of leptin relative to expenditure would cause an expression of PPAR-gamma and an increase in fat when relatively small amounts of food are consumed.

Leptin: Bridging the gap between the hCG protocol and maintaining thyroid function.

Leptin's elevation and depletion in the brain signals a fed or starved state, not only through hunger, but also through the metabolic suppression or stimulus from the thyroid. [8] When elevated, leptin stimulates thyrotropin-releasing hormone (TRH) that controls the release of thyroid stimulating hormone (TSH). TSH acts on receptors in the thyroid to promote synthesis and release of the thyroid hormones (T3 and T4), which increases the body's basal metabolic rate. [8] As blood glucose levels fall with very low-calorie diets, the depletion of leptin in the brain inhibits this cascade affect, resulting in a weaker metabolic signal from the thyroid. [9]

The natural drop in thyroid signal is an essential, life-sustaining mechanism that occurs during starvation. This mechanism slows down the rate at which the body needs fuel, thus preserving energy stores and life for a longer period of time. However, when leptin is administered during induced starvation, the thyroid signals stay strong. [10] If the thyroid signal stays strong, the body maintains a normal basal metabolic rate, and requires the same amount of fuel as if in a fed state.

To counteract the natural metabolic suppression of the thyroid, an energy-preserving survival mechanism, during sustained very low-calorie diets, an alternative stimulus of leptin would be needed.

Bridging the gap between hCG and leptin.

Not only is there clear evidence that the placenta produces leptin, but there's evidence that hCG might exert a negative feedback loop on trophoblastic release of leptin. [17][18][19][20] This means that specific quantities of hCG stimulate leptin production. If there were too much hCG, leptin levels would decline. If there were too little hCG, not enough leptin would be produced.

Based on these findings, hCG could be a viable stimulant of leptin. But the question yet to be answered is, would injections of hCG with Simeons' protocol:

- Stimulate sufficient blood leptin levels to interact with its receptors in the participant's central nervous system, acting as a potent FAS inhibitor in the hypothalamus, reducing hunger? In the hypothalamus to sufficiently stimulate the thyroid?

- Stimulate enough leptin production in the skeletal muscle to increase fatty acid β-oxidation? The right amount of leptin in the fat cells to prevent fat storage? And enough time and constraint for mitochondrial biogenesis to significantly increase resting energy expenditure?

- Could other forms of hCG administration, including homeopathic hCG, do the same?

Simeons' findings left many questions unanswered, and science uncharted by the refuting research of his protocol. Based on the evidence I've presented, hCG isn't burning fat, it doesn't directly reduce the appetite, and it doesn't stimulate the metabolism. *Rather, it is the energy demand of the protocol, combined with how hCG influences leptin, in the brain and body, that allows the body to metabolize fat and function as if fed, when food is not available.*

Chapter 16

A User-Friendly Explanation

Dr. Simeons was the doctor who created the hCG protocol. In the late 1920s, he studied the effects of *human chorionic gonadotropin* (hCG—a placental peptide hormone produced and released during pregnancy), as a means of therapy for obesity. In 1967, due to the demand for more information from both physicians and participants, Dr. Simeons privately wrote and self-published his observations, hypothesis, and a description of his protocol in *Pounds & Inches: A New Approach to Obesity*.

This manuscript describes how through years of deduction and observation, Dr. Simeons created a two-phase protocol. He believed that:

- A two-phase protocol properly adjusted the brain's hypothalamic disorder, which he considered as the root cause of obesity.

- The very low-calorie protocol, combined with hCG, would unlock the hypothalamus, allowing the body to sufficiently use stored body fat without having to force the breakdown of lean tissue as energy.

- The second phase would allow time for the previously "off-kilter" regulation of energy balance to realign a more balanced metabolic system. This would ultimately make the participant less susceptible to fat gain when compared to their pre-disposition to obesity before doing the hCG therapy.

Because Simeons' theories are based on the available information and science dated before 1970, his theories fall short of modern science. He also did not explain or prove his hypothesis through proper laboratory investigation, which is why the refuting research argued only on his observations—not his theories—focusing on the hypothalamic control of energy balance.

Today both Simeons' observations, as well as the refuting research, are outdated and lack the pertinent science required to prove it with a modern hypothesis and laboratory research. For this reason, we will not discuss his theories, but rather a modern hypothesis that could better explain how hCG could influence obesity, and prevent the symptoms of starvation during the first and second phase.

Fat is an organ.

Leptin is primarily a fat-derived hormone. Since its discovery in 1994, the sciences of starvation, feeding, how the body uses fat for fuel, and obesity, have exploded. Now that we know fat is an organ that can grow and deplete in significant size and cell count, our knowledge of how fat hormones function and influence other organs has completely evolved.

Today we understand leptin is a hormone that allows fat to be used for fuel, and also transmits a message to the brain and throughout the body—that a person has sufficient fuel. As food enters the mouth and stomach, leptin is secreted into the blood stream. Although leptin comes from many areas of the body, the most significant amount comes from fat.

Because some people have more fat than others, this source of leptin can significantly influence flux in blood leptin levels, which causes other organs to over- or under-compensate. Excessive or insufficient amounts of body fat can directly influence the balance point of the endocrine system, and over time, can re-set the metabolic "set point" of the body. This incredible ability to adapt within the brain and organs, could present symptoms of organ damage, signs of physical distress, and disorder within the endocrine system.

The intense focus on leptin, fat, and our organ's hormonal reaction, has been the center of attention. Today science can now explain with confidence leptin's role in sleep, ovulation, pregnancy, cardiovascular disease, Type II diabetes, Polycystic Ovarian Syndrome (PCOS), and more. To stay focused on what is significant to Dr. Simeons' hCG protocol, we will discuss only leptin's role with hunger, fat metabolism, fat gain, and the thyroid.

To start, let's review the most important questions that participants, practitioners, and skeptics need to discuss:

1. How can an adult continue to eat only 500 calories for weeks on end, and not feel hungry?

2. How could energy and metabolism stemming from the thyroid not decline during the very low-calorie protocol?

3. With such a small amount of food, how could body fat sufficiently provide fuel without having to rely on lean tissue as back up?

With today's modern science, all three questions can be answered and proven with laboratory research.

Leptin: hunger, fat metabolism, and the thyroid.

As blood leptin levels rise, and leptin increases in the brain, the hunger center of the brain deactivates, communicating to the conscious mind you have adequate fuel, and diminishing the urgency to eat. On the other hand, if leptin levels decline, the same hunger center senses a decline in fuel, and activates and communicates to the conscious mind an increased urgency to eat. Remember that leptin only removes the urge to eat; it *does not* create fullness.

$$\uparrow \text{eating} = \uparrow \text{leptin} = \downarrow \text{hunger}$$

If you were to go on a 500-calorie diet, there wouldn't be adequate leptin released from fat and other areas of the body to sufficiently deactivate this center of the brain. An obvious increase in the urge to eat would occur.

$$\downarrow \text{eating} = \downarrow \text{leptin} = \uparrow \text{hunger}$$

The difficulty here is in the interpretation of *hunger*. Hunger is too subjective in humans. Many people eat without hunger, which can cause problems with this hormonal system. Take for example the difference in hunger between a person who has very little fat, compared to an individual with a very large amount of fat. Remember, the most influential source of blood leptin comes from fat, so how much fat a person has will absolutely affect how often they experience hunger, and how much food is necessary to alleviate the urge to eat.

For example, a lean person might need to eat two apples to stimulate enough leptin from his or her small amount of fat to remove hunger in the brain. In comparison, an obese individual might only need half an apple to get the same amount of leptin, and the same relief from hunger.

↑ fat cells = ↑ leptin =
↓ need for food

↓ fat cells = ↓ leptin =
↑ need for food

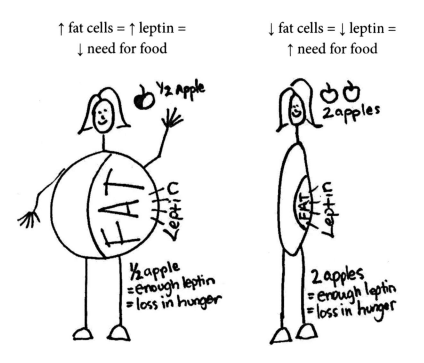

Leptin signals to the brain that the body has been sufficiently fed by alleviating hunger, but if the food that is eaten isn't digested, how could it possibly provide immediate fuel? What "feeds" the body while the food is still in digestion? To answer this, we need to discuss what leptin does to allow stored fat to be used for energy as the eaten food is digested.

As blood leptin levels rise while eating, fat cells begin metabolizing fat for fuel. Leptin opens access for fat to enter into the cells mitochondria. The mitochondria is an organelle in each fat cell where the fat gets chewed up and spit out into the body as human "gasoline" for muscles and organs. This immediate fuel availability energizes the functions of the body, and preserves blood glucose to be used in the brain.

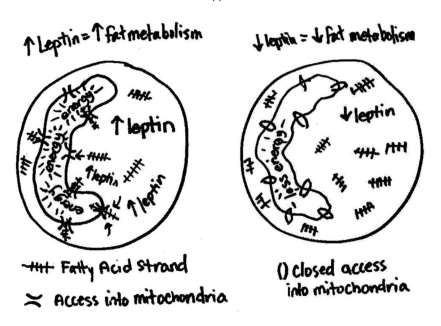

↑Leptin = ↑fat metabolism

↓leptin = ↓fat metabolism

↑leptin

↓leptin

―#tt― Fatty Acid Strand

)(Access into mitochondria

( ) closed access into mitochondria

The immediate availability and conversion of fat into fuel provides what the body needs while food is digested.

- Could too much fat create a disproportionate breakdown of fuel, creating more than what's necessary?

- Could this create a backup or stockpile of fuel, if it isn't used at the rate it's being produced?

- If so, could this cause a rise in blood glucose that would warrant the need for insulin to recreate fat out of the energy?

Unfortunately, large amounts of leptin seen in obese individuals could be a sign they eat far more than their body needs, producing too much leptin relative to the fuel they need. In fact, the more leptin an individual produces, the more fat fuel they create, and without equal demand, this causes an increase in the hormone PPAR-gamma that controls fat droplet formation and new fat cell maturation. These new fat cells show up in areas you didn't have fat before, and are very difficult to lose. These areas of fat are highly associated to disease such as Type II diabetes and heart disease. But without these new fat cells, you would not have enough fat to store the excessive production of fuel. This fat-gaining response aids both insulin and the liver by creating new and bigger fat cells for adequate storage.

But these new fat cells can cause an even bigger problem because they can produce more leptin than normal-sized fat cells. This means they create more leptin when stimulated, and provide more fuel for the body. Each new larger fat cell magnifies the leptin-induced problem, and could cause excessive fat-fuel production without warrant.

For example, compare a person who has only 10 lbs. of fat to a person who has 100 lbs. of fat. When they eat the same exact food (as food stimulates leptin), each cell of fat is stimulated to produce leptin, and then fuel. Thus the person with more fat stimulates more leptin, and consequently more fuel than the person who has less fat cells. This is why people who are very low in body fat tend to be hungrier, and can also eat more without gaining fat, and why people who have too much fat are minimally hungry, and gain fat even when they eat very little.

Each fat cell magnifies the function of this fueling system, so the amount of fuel a person gets is not measured by the food they eat, but rather by the amount of fuel their fat cells provide. So two people of different body fat can eat the same amount of food—and have a completely different fueling response. Simply put, the more fat cells you have, the less food you need to get appropriate amounts of leptin. The less fat you have, the more food it takes to simulate enough leptin to acquire enough fuel.

If an obese person eats two apples, his or her fat will provide fuel that is more than what was in the two apples she or he ate. This over-produced fuel requires a faster metabolism to balance fueling input with output, without gaining more fat.

↑ fat + 2 apples = ↑ leptin =
↑ fat gain

↓ fat + 2 apples =
energy balance

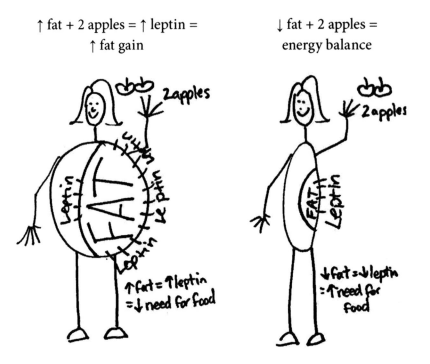

The amount a person should eat is not relative to how many calories they burn, but rather relative to how much fat they have, and how much leptin they produce. Remember, fat isn't just a stockpile of energy, it is an organ that also coordinates with other organs. So if you have too much "fat organ," the other organs are definitely going to have to compensate. That is, unless the one conscious signal is used when leptin is too low or just right: hunger.

The convenient part about leptin is that when it declines, hunger is the result, and it's easy to know when eating is appropriate. Food and hunger is a perfect match because hunger results from lowering leptin levels, and food stimulates the body to make more. A person can individualize how much he should eat by listening for this signal, waiting for hunger to present itself, and eating only enough for this hunger to confidently be removed.

With the prior example, if the obese individual had listened, she or he would have noticed his or her hunger diminished after eating one half of the first apple. However, the leaner person with much less fat, needed to eat two apples to stimulate enough leptin to feel relief from hunger.

Simply put, if you're not hungry, your leptin levels are elevated, and your body is using stored fat as energy. If physical hunger increases, and the urge to eat is felt, then leptin levels are dropping, fat is insufficiently fueling the body, and the need for food is higher.

The body's response to starvation is triggered by the drop in blood leptin levels, which not only results in hunger, but also the suppression of the thyroid. As food is eaten, leptin stimulates the center of the hypothalamus that in turn controls the intensity of the thyroid signal. With a 500-calorie diet, when blood leptin levels fall, it would be appropriate to assume the thyroid signal would also weaken. This is a life-saving mechanism that weakens the body's energy system to sustain at a lower level, if fuel is not sufficiently supplied by reserves. This mechanism is like a light bulb that brightens or dims, using a dimmer switch. Leptin is the hormone that triggers the thyroid dimmer switch in the brain to turn up the signal, and energize the body.

↑ food consumption = ↑ leptin = ↑ thyroid signal = ↑ metabolic rate

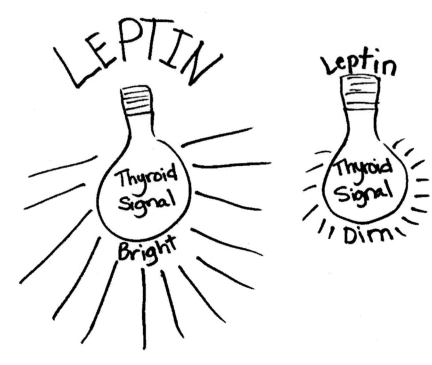

↓ food consumption = ↓ leptin = ↓ thyroid signal = ↓ metabolic rate

Not only does leptin influence hunger and fat metabolism, it also affects the intensity of the thyroid signal, which controls the basal metabolic reaction of the body. Because leptin indicates to the brain and body there is sufficient fuel, a lack of leptin would signal insufficient fuel, consequently triggering the energy and fuel-conserving response to starvation. To better understand how Dr. Simeons' 500-calorie protocol prevents such a life-preserving response, we must understand how hCG influences the signal of leptin.

In a pregnant woman, the placenta produces its own leptin, ensuring a minimum "fed" state for the baby. Laboratory research has discovered that hCG and leptin have a "negative feedback loop." This means if hCG is too low in the placenta, leptin increases and stimulates more hCG production. As hCG reaches a level that is adequate, leptin stimulation decreases until the amount of hCG once again drops to an insufficient amount.

In America, pregnant women gain 15% body fat on average, and it could be attributed to the extra supply of leptin from the placenta. Other hormones increase blood leptin levels that break down fat storage for energy, and reduce hunger, even if food isn't eaten. If food is eaten anyway, the obvious outcome is increased fat storage and new fat cells. On the other hand, if food isn't available, the stimulus of leptin from other hormones, such as hCG, signal a "fed" state, making available fat stores to sufficiently supply energy for both the mother and the baby. This sensitized mechanism gives the baby a baseline insurance of energy, making the mother's fat-burning capacity more available, but also magnifying her fat-storage capacity.

Unfortunately, if a woman gains significant amounts of fat during pregnancy, especially new fat cells formed from too much leptin, her affinity to produce too much leptin after delivery will go up. This might cause the entire endocrine system to shift in a way that could cause long-term symptoms and sensitivity to fat gain.

It's not uncommon for women to feel they are more sensitive to fat gain after having children, and to also have symptoms that are caused from the compensatory response of an under- or over-active organ, such as the thyroid. Some lose fat cells during pregnancy, and notice it's more difficult to gain fat than before, and have lost symptoms that could have stemmed from other organs. If the metabolic system can be so greatly influenced during pregnancy, with the gain or loss of fat, could we influence the same system by injecting the same pregnancy hormones that effect leptin, while manipulating food intake to cause increased demand on stored fat? Based on what Simeons observed, and what participants experience who follow the protocol meticulously, this could be true.

If small amounts of hCG stimulate leptin, proper laboratory experimentation and research could prove whether or not hCG can stimulate enough blood leptin levels to reduce hunger, sustain normal thyroid signal, and optimize fat metabolism to sufficiently fuel the body. Research could also show that different methods of administration and amounts of hCG might be appropriate for some who are more obese, and different for those who are less so.

If hCG does stimulate leptin from fat stores, the requirement for food to stimulate leptin would be drastically reduced, and eating wouldn't be as vitally important from a fueling standpoint. But if food is overeaten, along with the extra stimulus of leptin from the administration of hCG, fat gain and other organ compensations would be the obvious result. This could be why 500 calories of controlled food might actually be appropriate, while even less food is necessary for those who have more fat.

With the increased demand on fat to create fuel for the body, the obvious result would be the removal and loss of excessive fat organ cells. As each fat cell depletes, less leptin would be stimulated, and the entire organ system of the body would be affected. Organs that were hyperstimulated by leptin would get relief. Organs that were suppressed by too much leptin are now reactivated as leptin levels become balanced. With time and consistency, the entire hormonal system of the body would acclimate and create a new balance, optimally working with less leptin. The obvious goal would be to reduce fat organ cells to the appropriate amount that works optimally with the rest of the organs in the body. This would effectively reduce risks of death and disease.

The hCG protocol could be a legitimate therapy for the entire hormonal regulation of the body that was negatively influenced by excessive production of leptin. For this reason, the hCG protocol should not be approached like a diet. Without the strict restriction in food intake, leptin would not be controlled, demand would not be created, and more leptin-induced harm could be the consequence.

What we are dealing with is the hormonal system of the body, which is the entire endocrine system including the regulatory functions from within the brain. Dr. Simeons did not create this method as a diet, but as a true medical therapy for obesity and resulting life-threatening diseases.

With that said, do not attempt the hCG protocol unless you are ready to let go of food emotionally, no matter what the weight-loss result.

Protocol: 125 IU's hCG = ↑ leptin + VLCP = ↑ fat loss

After Protocol: ↓ fat = ↓ leptin + ↑ food intake = endocrine balance

Chapter 17

What to Expect—and What Most Experience

> *"In dealing with a disorder in which the patient must take
> an active part in the treatment, it is, I believe, essential
> that he or she have an understanding of what is being
> done and why. Only then can there be intelligent coopera-
> tion between physician and patient."*

- A.T.W. Simeons, *Pounds & Inches*

Understanding the key roles hormones play in hunger, satiation, energy balance, and metabolism, makes the protocol less mysterious. It helps explain why Simeons required "loading," the specificity of the food to be consumed, and why the maintenance phase is just as important as the first phase of the protocol.

Unless hCG stimulates enough blood leptin to register a "fed" state in the brain and body, the very low-calorie protocol would immediately induce symptoms of starvation. Research has shown that the treatment of recombinant hCG in the placenta revealed a stimulatory effect on endogenous leptin expression, with the maximal effect at the amount of 100 IU/ml. [18] This is remarkably close to the amount of hCG Dr. Simeons prescribed (125 IU/ml). The question is how much hCG would be ideal to adequately stimulate leptin from fat? Also, would the first injections, or any other form of hCG administration, instantaneously stimulate adequate blood leptin? Probably not, which might be why Simeons required "loading."

Dr. Simeons observed a required three days of injections before starting the VLCP. He claimed the first three injections were "non-effective," and during these days, forced feeding of fatty foods simultaneously stocked normal fat reserves in preparation for use of abnormal fat. As he describes force-feeding in *Pounds & Inches,*

> "...One cannot keep a patient comfortably on 500 calories unless his normal fat reserves are reasonably well stocked. It is for this reason also that every case, even those that are actually gaining must eat to capacity of the most fattening food they can get down until they have had the third injection. It is a fundamental mistake to put a patient on 5oo calories as soon as the injections are started, as it seems to take about three injections before abnormally deposited fat begins to circulate and thus become available."

Loading might be an essential first step, connecting the relationship between the placental hCG administered with your fat cells to stimulate adequate leptin. Without loading, it could take two weeks before symptoms of starvation diminish. Many people who've attempted the protocol without taking the hCG while loading, who don't eat enough, or who don't prioritize fat during the loading, often experience more hunger during the first and second week. These people tend to falsely assume the protocol just doesn't work, when in reality they weren't instructed how to load properly, didn't understand the instructions, or didn't take the loading seriously. Lack of compliance is what caused the problem.

New research could explain the relationship between loading and hunger during the VLCP, and through proper instruction, increase participant compliance for favorable hunger results. Without scientific explanation, many participants find loading mysterious, which makes it easier to assume it isn't important. Actually, it could be the most vital step necessary for hCG to have the proper effect on leptin before restricting consumption to the 500-calorie protocol.

However, once the hCG stimulates sufficient blood leptin, the protocol constraints would require meticulous control not to over- or under-stimulate blood leptin and blood glucose. This counter action begins with the specificity in what, how much, and when food is eaten. The foods listed in Dr. Simeons' *Pounds and Inches* are very simple. Most likely they work because they stimulate leptin minimally, but provide maximal nutritional value. The most important food to eat at minimum, is the protein. The body can't create all the amino acids necessary, so an outside source is needed to prevent protein deficiency.

Without the specific control of food intake, too much leptin could be stimulated, PPAR-gamma activated, and a high susceptibility for fat gain would be expected. This could be why pregnant women typically gain up to 15% more body fat during pregnancy.[29] This also might explain why Simeons observed significant sensitivity to food, and why it took him decades to construct the protocol's specificity and rigidity to accurately control the physical environment.

Many who justified wavering from the mandated food restrictions experienced a weight gain of 2-3 pounds, following a minor "cheat." If cheating is done with more protocol-approved food, you may still observe gains. Whether this gain is fluid shifting, waste, or fat, we won't fully understand until research tests protocol food, leptin, insulin, and the body's sensitivity to gain during hCG administration.

With predictable and consistent administration of appropriate amounts of hCG, and the rigid control of food consumption, the next biggest influences to energy balance come from hormonal influence and expenditure.

Hormones that directly influence leptin, such as insulin, estrogen, progesterone, and more, could cause fluctuations in the protocol's balancing act. Hormone therapy, medications, and even lotions and oils in makeup, were observed to sway results during the protocol.

Simeons wrote in *Pounds & Inches*:

> *"Most women find it hard to believe that fats, oils, creams and ointments applied to the skin are absorbed and interfere with weight reduction by hCG just as if they had been eaten. This almost incredible sensitivity to even such very minor increases in nutritional intake is a peculiar feature of the hCG method. For instance, we find that persons who habitually handle fats, such as workers in beauty parlors, masseurs, butchers, etc. never show what we consider a satisfactory loss of weight unless they can avoid fat coming into contact with their skin."*

This hormonal sensitivity could be why Simeons observed changes during female menstruation, and why he suggested stopping injections, but not the VLCP, during this time. Leptin not only has a negative feedback loop with hCG, but it increases the plasma levels of luteinizing hormone (LH) and follicle stimulating hormone (FSH). [20] Also, this hormone-sensitive leptin stimulus might explain why women going through menopause experience a propensity to gain fat, even with limited food. So in both men and women, ideally outside hormonal influence should be minimized to consistently sustain an optimized amount of leptin. Another influence that can throw off this balance is physical activity.

Participants could justify additional laborious exercise with the misdirected belief that results are a linear consequence of caloric expenditure. Due to the nature of how the protocol might work, if the host of enzymes and hormones involved cannot balance the expenditure, fat metabolism could be too slow and insufficient to meet demand, causing other tissue substrates to fulfill the gap. However, the physical sensitivity to both expenditure, as well as hormonal influence, is a result relative to body-fat composition.

Body-fat composition has a large impact on the hormonal environment of the protocol because leptin's source primarily comes from fat. Research has shown that the more fat a person has, the more leptin is produced. [21] During the protocol, more obese participants might show more sensitivity to the influences of food and hormones, and while leaner participants might be more sensitive to expenditure.

To start, those who have more body fat likely need less protocol food to feel satiated, and have fewer hunger fluctuations to create the ideal energy homeostasis. More obese clients could also be more sensitive to oils in lotion and makeup, as well as to diet drinks or anything that could stimulate leptin. For this reason, without strict protocol compliance, they might experience a higher sensitivity to fat gain.

Leaner participants show less sensitivity to hormonal influences, and less affinity to gain weight, with minor adjustments. Leaner participants could also show more sensitivity to increased activity, and more fluctuations in hunger.

Many participants find that as they lose weight, their hunger increases. When fat is reduced to an amount that insufficiently stimulates blood leptin, the transmission of leptin through the CNS, might not be high enough to counteract a decrease in hypothalamic malonyl-CoA. Again, this reduction of the hormone in the brain increases the expression of NPY and AgRP, increasing the physical urge to eat, and the sensation of hunger. This information is explained in chapter 15.

More obese participants seem to lose more fat per day than their leaner counterparts. Dr. Simeons predicted with adherence that participants would lose a pound of fat a day. However, fat loss would only be predictable if there was a controlled blood-leptin stimulus, consistent resting energy expenditure, and minimal hormonal influence. Only then would a participant burn fat at a rate that matches his or her energy demand. If that demand is 3,500 calories above the energy supplied by food, only then would the loss of a pound of fat lost per day make sense. A person isn't going to burn a pound of fat a day if he or she doesn't create the demand for that much energy, so smaller and leaner participants tend to lose less fat. More obese participants typically burn more fat in the same period of time because of the extra energy demand imposed by their weight against gravity.

Which fat stores are utilized, and where fat is lost during the first phase of the protocol, could be observed due to the areas of fat that stimulate more leptin than others, [5] making these fuel sites more accessible during the protocol's imposed energy demands, but more susceptible to gain with any diversion from the regime. Before the protocol, these areas were more likely to be sensitive to gain, especially if they produce more leptin than other areas. With regular diet restriction, these leptin-sensitive fat stores might also be sensitive to ACC, thus increasing muscle malonyl-CoA, and preserving these fat stores (as discussed in chapter 15). This perhaps explains Dr. Simeons' observation of "abnormal fat." The more of this leptin-sensitive fat a participant has, the more leptin is produced, making this fat more sensitive to gain.

But regardless of whether hCG is or isn't present, fat will be lost during a 500-calorie diet. The difference is that hCG's influence on leptin allows more fat loss, and prevents lean tissue loss, making body fat composition improvements more drastic than those without hCG. This is not measured by weight alone, but with specific measurements that can assign a percentage of body fat to the body's total weight.

Modern technology and improved assessments of body-fat composition more easily assess the results of weight loss, as well as maintenance of composition, through the second phase. New observations could essentially remove doubt as to whether weight gain is from fluid shifting, constipation, or confirm if it's from fat.

For weeks, and for some, close to two months, participants rigidly measure their food, and constantly weigh themselves. As the first phase comes to a close, and the second phase begins, many aren't prepared for the increased flexibility and less-rigid constraints. Without proper instruction, participants often cause physical harm and stress to their digestive system when immediately adding too much food, with too much fat. Pancreatitis, dumping syndrome, gall bladder inflammation, and flu-like symptoms, are not uncommon consequences for participants who go from the extreme rigidity to comparative freedom.

After almost two months of the first phase, the body could be hypersensitive to producing leptin. Slowly alleviating hunger by strategically adding foods that are less potent often is a vital regulatory step after the controlled environment of the first phase is over. Also, the rest period might be necessary for mitochondrial adaptation, potentially elevating resting energy expenditure as time permits (explained in chapter 15).

Resting energy expenditure would also be improved due to a more moderate stimulus of leptin, and the physiological reaction required to find a new homeostasis, as a result. This would allow the complex system of tissues and organs to function appropriately with the regulation from the brain, and the outside stimulus of both food intake and expenditure. The protocol could be a miraculous therapy to rebalance a resting metabolic set point that was offset as a response to abnormally high secretions of leptin. As Simeons stated in *Pounds & Inches*:

> *"In treating obesity with the HCG + diet method we are handling what is perhaps the most complex organ in the human body. The diencephalon's functional equilibrium is delicately poised, so that whatever happens in one part has repercussions in others. In obesity this balance is out of kilter and can only be restored if the technique I am about to describe is followed implicitly. Even seemingly insignificant deviations, particularly those that seem to be an improvement are very liable to produce*

most disappointing results and even annul the effect completely."

Unfortunately, because most participants and practitioners know very little about the new science of leptin and its role in starvation, the protocol is still misunderstood, and misdirected, and many have attempted to "socialize" it, to make it appeal to the masses.

New methods, with more food, are now being introduced—and not because of proper clinical research, but marketed to ease the fears of those who've yet to fully understand how the protocol works. Until there is sound laboratory research, clinical evaluations, and conclusive evidence, we are all speculating—some more intelligently than others.

> *"I must beg those trying the method for the first time to adhere very strictly to the technique and the interpretations here outlined and thus treat a few hundred cases before embarking on experiments of their own, and until then refrain from introducing innovations, however thrilling they may seem. In a new method, innovations or departures from the original technique can only be usefully evaluated against a substantial background of experience with what is at the moment the orthodox procedure."*

-Dr. A.T.W. Simeons, *Pounds & Inches*

Prescription vs. Homeopathic hCG

Not only has there been an increase in hCG protocol prescriptions, but a huge influx of homeopathic hCG into the weight-loss market. Because homeopathic hCG only contains trace amounts of hCG, the FDA doesn't require a prescription for its use. It's much less expensive, easier for the consumer to obtain, and is less burdensome for businesses to produce.

The debate within the hCG weight-loss industry is whether oral or topical hCG, and homeopathic hCG, are just as effective as hCG injections.

The only way to prove whether one is as effective as the other would be to measure blood leptin levels, not blood hCG levels.

Although other forms of hCG administration might stimulate leptin production, there is less control and predictability compared to injections. This lack of control could cause more erratic hunger, less consistent fat loss, and other leptin-induced (or lack thereof) symptoms. With less leptin control, less post-protocol metabolic increase would occur. We won't fully understand it until we have concrete evidence of whether administered hCG, combined with the protocol, actually maintains blood leptin levels sufficient to sustain healthy energy homeostasis.

The hCG protocol has become an overnight sensation without proper application. Participants, and the businesses that provide the protocol, have relied on outdated and disputed theories, creating a tendency to inappropriately approach the protocol as if it is a diet.

The hCG protocol is not a diet. Simeons meant for it to be a medical rehabilitation for symptoms and diseases associated with obesity. Now, without standards or validity, its uncontrolled and improper use is spreading like wildfire. As effective as the protocol might be, ignorance about how it should be administered and adhered to, causes people to change it, cheat on it, and become frustrated with it when their results are unsatisfying.

> *"Any patient who thinks he can reduce by taking a few 'shots' and eating less is not only sure to be disappointed but may be heading for serious trouble. The benefit the patient can derive from reading this part of the book is a fuller realization of how very important it is for him to follow to the letter his physician's instructions."*
>
> - Dr. A.T.W. Simeons, *Pounds & Inches*

Chapter 18

Observations Based on Our Results

The first two years I helped people with Dr. Simeons' protocol, I recorded observations to a battery of tests for about 300 participants. These observations should give additional insight for those who prescribe hCG for Simeons' hCG protocol. All these observations were with participants who used injections of hCG; no orally administered hCG was used.

Here's what's been tested and observed:

Resting Energy Expenditure (REE): Indirect Calorimeter

- Using the Korr Ree Vue indirect calorimeter (model 8100), we tested participants immediately before the first injection, and three weeks after the final injection; each participant was given a total of 23-40 injections, at 125 IU, as Simeons suggested.

- Participants show an average increase in resting energy expenditure of .98 calories/lb/day.

- Subjects who were higher in body fat composition, and injected longer, showed larger increases than those who had less fat, and shorter first phases.

- Overall, if the protocol was strictly followed, participants burned more calories per pound of body weight each day than they did before the protocol was administered.

Blood Pressure

- Each subject's blood pressure was recorded before the protocol began, and checked weekly throughout the first and second phase of administering the protocol.

- Average blood pressures before protocols were 127/80.

- During the VLCP, blood pressured dropped to an average of 115/75. After finishing both first and second phases, blood pressure rose to 119/75.

- Overall, both systolic and diastolic blood pressures were lower than before protocol participation.

Resting Heart Rate

- Each subject's resting heart rate was measured before the protocol began, and checked weekly throughout the first and second phase of administering the protocol.

- Overall, resting heart rates decreased. Participants with considerably more body fat had more significant decreases in resting heart rate, for some, they fell below 60 beats per minute during the VLCP.

- The average drop was from 75 (bpm) before the protocol to 69 (bpm) after both first and second phase of the protocol.

Body Fat Percentages

- Each subject's body fat composition was measured with calipers using an average between the Durnin/Womersley 4-skinfold formula and Jackson/Pollock 3-skinfold formula.

- We measured before the protocol began, and weekly throughout the first and second phase of administering the protocol.

- We observed a 1% average drop in total body fat in each subject per week, with clients who rigidly adhered to the protocol.

- Those who did not follow the protocol showed a slowing of fat loss and/or maintenance of body fat percentage.

- Pounds lost per week paralleled what is metabolically predicted from the pre-protocol REE test, when the protocol was strictly followed.

Circumference Measurements

- Fat loss was observable in almost all parts of the body, but was most noticeable in the visceral areas of the body: abdomen, rib, thigh, and back fat reductions were most common.

- Some subjects experienced losses of 5 inches of circumference in one week during the protocol, from one abdominal measurement.

- It was more common to see 2-3 inch drops per week.

Treadmill Cardiovascular Test

- We used a five-minute walking cardiovascular test, measured with heart rate monitors, and with the treadmill incline progressively increased.

- The test was administered before the protocol began, and three weeks after the last injection. Most subjects showed reduced heart rates.

- The majority showing decreases between 10 and 20 beats per minute, when compared to their heart rates before the protocol.

- Some participants showed heart rate reductions of over 40 beats per minute, even with only 20 pounds of weight loss.

Flexibility

- We measured modified flexibility using a sit-and-reach test before the protocol began and three weeks after the last injection.

- All tests subjects showed an increase in hip and spine flexibility, and some had gained up to 4 inches in increased reach.

Strength Test

- Each subject was evaluated before the protocol began and three weeks after the last injection was completed, with a standard 10 rep max bench press test.

- The majority of tests showed maintenance without measurable strength increase.

Strength Endurance

- Each subject was evaluated for endurance before the protocol began, and three weeks after the last injection was completed, using a one-minute push-up and sit-up test.

- All subjects showed marked increases in completed repetitions.

Preliminary Statistical Analysis (for the first 40 participants)

This preliminary analysis was done by Lee Hannah of Boise State University-Boise, Idaho.

Lee Hannah, DVM, MS, MPH
Assistant Professor, Medical Epidemiologist, Boise State University

The data set is small, containing only 12 variables. Of interest was the person's gender (only 4 males to 36 females), weight at the beginning of the hCG protocol, resting energy expenditure(REE) at the beginning of the protocol, calorie/lb/day at the beginning of the protocol, weight at the end of the protocol, REE at the end of the protocol, and calorie/lb/day at the end of the protocol.

Because we have pre- and post-information on the same individuals, I used a paired t-test to look for a significant change from post to pre for both weight and REE.

Weight: For weight, there was a significant reduction in weight across the protocol. On average, across all 40 individuals, weight went from 202.73 pounds to 180.35 pounds (a change of 22.38 pounds). This was statistically significant, with a p-value <0.001.

REE and calorie/lb/day: REE also dropped between the pre and post time periods, with an average reduction of 69.48; however, this difference only reached borderline significance, with a p-value = 0.07. What is more important to the study is the fact that the calorie/lb/day (which is calculated as the REE divided by body weight) increased across the 40 participants. At baseline, the participants were burning 9.51 calorie/lb of body weight and at completion of the protocol the average participant was burning 10.30 calorie/lb of body weight. This was statistically significant, with a p-value <0.001.

Table 1: Paired Samples Statistics

		Mean	N	Std. Deviation	Std. Error Mean
Pair 1	post weight	180.35	40	48.085	7.603
	pre weight	202.73	40	55.404	8.760
Pair 2	Post REE	1828.40	40	451.720	71.423
	Pre REE	1897.88	40	413.026	65.305
Pair 3	calorie/lb/day2	10.3013	40	1.58841	.25115
	calorie/lb/day1	9.5094	40	1.36931	.21651

Table 2: Statistical significance of differences observed

		Paired Differences		95% Confidence Interval of the Difference		
		Mean	Std. Deviation	Lower	Upper	Sig. (2-tailed)
Pair 1	post weight - pre weight	-22.375	9.953	-25.558	-19.192	.000
Pair 2	Post REE - Pre REE	-69.475	232.489	-143.829	4.879	.066
Pair 3	calorie/lb/day2 - calorie/lb/day1	.79190	1.23300	.39757	1.18623	.000

These results were achieved with test subjects who were limited to eating fewer than 500 calories/day, sustained from four to six weeks.

By providing this data, I don't intend to convince anybody of the efficacy of the hCG protocol, but to confirm what others have observed, and to also motivate more data collection by those who are on the front lines of observation. More importantly, the goal is to discuss a new hypothesis that would motivate new research and more specific data collection from those who are medically and scientifically qualified to collect and analyze it properly. For example, because we used injections of hCG, we do not know whether the use of oral hCG would have the same physical results. More research is desperately needed.

Chapter 19

How the Protocol Is Done

DR. SIMEONS' HCG PROTOCOL									
The First Phase							The Second Phase		
The first three days of hCG administration require a minimum of two days of forced feeding. Eat high fat foods all day long. Do not allow hunger on these days.	Week 1	Week 2	Week 3 (minimum)	Week 4	Week 5	Week 6 (maximum)	3 day transition (no hCG taken)	Second Phase Week 1	Second Phase Week 2	Second Phase Week 3
LOADING: Days 1-3	THE VERY LOW CALORIE PROTOCOL: 3-6 Weeks							LOW CARBOHYDRATE PROTOCOL: 3 weeks		

Dr. Simeons' hCG protocol includes two phases. The first phase includes a three-day loading process, the very low-calorie protocol (VLCP), and a transitional process into the second phase. The second phase is the vital step requiring consistent hormonal balance, which allows for the entire endocrine system to recalibrate a new equilibrium or "set-point." Because a significant amount of body fat was lost during the first phase, the brain and other organs must readjust their coordinating system to match having fewer fat hormones.

One could argue the second phase is as fundamentally important as the first phase. The VLCP removes excess fat organ tissue, and the second phase readjusts the response from the brain and other organs. However, if the first phase is not properly followed, the entire process could actually cause harm, and without proper hormonal stabilization during the second phase, the body might react by inducing a return in fat. For example, without loading, the participant has significant side effects due to starvation, and if he cheats, more fat gain could exacerbate leptin-induced organ response and disorder.

For this reason, it is very important participants fully respect the protocol as a therapy for the endocrine system, and have a thorough understanding of what the protocol entails, how it works, and why it is *not* a diet. If it is approached with the same loose adherence as with diets, the participant will surely be disappointed.

The first phase begins with the first injection of hCG, and ends with the last day of the 500-calorie protocol. The duration for most people lasts between 3-6 weeks. The beginning of the first phase starts with loading, sustains with the very low-calorie protocol (VLCP), and ends in transition before starting the second phase.

THE FIRST PHASE: Loading

To start the protocol, the relationship between the small dose of hCG and fat's stimulation of leptin requires three days of forced-feeding.

Eating high fat foods to full capacity during those days acts as a catalyst, ensuring sufficient blood leptin is stimulated by the fourth day that warrants the drastic drop to the 500-calorie protocol.

Loading is a fundamental necessity for the entire protocol to work. Without loading, participants can experience symptoms of starvation for longer than two weeks. Because of how relentless loading is, many people have difficulty force-feeding for all three days, and tend to load less on the third day. We suggest that participants be extremely diligent about loading on day two and day three. The goal during the second and third day is to eat as much fatty food as possible—without causing physical harm. It would be safer to eat continually all day, rather than mass amounts in one sitting.

Here is an example load day:

Breakfast:

- Whole cream in coffee, omelet with cheese and meat, extra bacon and sausage, croissants with butter on the side

Snack:

- Chips and guacamole

Lunch:

- Triple layer cheeseburger with curly fries, dipping sauce for fries, and a peanut butter milkshake

Snack:

- Potato chips with cheese dip

Dinner:

- Deep-dish pizza with all the meats and cheeses

Dessert:

- Cheese cake, extra whole whipped cream

Obviously you shouldn't eat so much that you vomit, or put yourself at risk of causing injury to your stomach. The food needs to be fat- and grease-laden for the hCG to properly affect the body's metabolic system.

Here is what Dr. Simeons says in his manuscript:

Gain before Loss

*Patients, whose general condition is low, owing to excessive previous dieting, must eat to capacity for about one week before starting treatment, regardless of how much weight they may gain in the process. One cannot keep a patient comfortably on 500 calories unless his normal fat reserves are reasonably well stocked. **It is for this reason also that every case, even those that are actually gaining, must eat to capacity of the most fattening food they can get down until they have had the third injection**. It is a fundamental mistake to put a patient on 500 calories as soon as the injections are started, as it seems to take about three injections before abnormally deposited fat begins to circulate and thus become available.*

We distinguish between the first three injections, which we call "non-effective," as far as the loss of weight is concerned, and the subsequent injections given while the patient is dieting, which we call "effective." The average loss of weight is calculated on the number of effective injections and from the weight reached on the day of the third injection, which may be well above what it was two days earlier when the first injection was given.

Most patients who have been struggling with diets for years, and know how rapidly they gain if they let themselves go, are very hard to convince of the absolute necessity of gorging for at least two days, and yet this must be insisted upon categorically if the further course of treatment is to run smoothly. Those patients who have to be put on forced-feeding for a week before starting the injections usually gain weight rapidly—four to six pounds in 24 hours is not unusual—but after a day or two, this rapid gain generally levels off. In any case, the whole gain is usually lost in the first 48 hours of dieting. It is necessary to proceed in this manner because the gain re-stocks the depleted normal reserves, whereas the subsequent loss is from the abnormal deposits only.

Patients in a satisfactory general condition, and those who have not just previously restricted their diet, start forced-feeding on the day of the first injection. Some patients say that they can no longer overeat because their stomach has shrunk after years of restrictions. While we know that no stomach ever shrinks, we compromise by insisting that they eat frequently of highly concentrated foods such as milk chocolate, pastries with whipped cream, sugar, fried meats (particularly pork), eggs and bacon, mayonnaise, bread with thick butter and jam, etc. The time and trouble spent on pressing this point upon incredulous or reluctant patients is always amply rewarded afterwards by the complete absence of those difficulties that patients who have disregarded these instructions are liable to experience.

During the two days of forced-feeding from the first to the third injection— many patients are surprised that contrary to their previous experience, they do not gain weight, and some even lose. The explanation is that in these cases, there is a compensatory flow of urine, which drains excessive water from the body. To some extent, this seems to be a direct action of hCG, but it may also be due to a higher protein intake, as we know that a protein-deficient diet makes the body retain water.

Forced-feeding of high-fat food might be a catalyst in connecting the relationship between hCG and leptin. Perhaps those who have less fat need to force-feed all three days, and many experience a couple of days of transitional hunger that usually subsides by the second or third day of the VLCP. *Do not* assume you will be getting a head start in the weight loss game by not loading. Without loading, the hCG will not work, and neither will the protocol.

The First Phase: Very Low-Calorie Protocol (VLCP)

The very low-calorie protocol (500 calories or less) starts on day four, and lasts a minimum of 20 days and a maximum recommendation of around 40 days. This is the most miraculous part of the protocol, and contrary to what would happen without hCG, leptin levels should remain elevated, allowing fat metabolism to feed the body, preserving organs and muscles, and fueling the brain and body with minimal food and manageable hunger.

Because of hCG's stimulus of leptin, you should have ample fuel supplied by fat, which makes food energy much less necessary, and prevents blood glucose levels from dipping too low. If the protocol restrictions are not followed, you have a high probability of fat gain as leptin is over-stimulated, causing excessive fat breakdown and a surplus of fuel, along with the subsequent stimulus of PPAR-gamma, which triggers the formation of fat droplets and new fat cell formation. This is why the VLCP is imperative after the loading process, and why this is a protocol or procedure, *rather than* a diet.

Keep in mind: hCG is a hormone that influences the hormonal system of the body. Do not approach the VLCP as if it is a diet that can be followed loosely. Without strict adherence, stress will be inflicted on the entire organ system of the body, including the brain.

Here is how Simeons described the VLCP that he created after many years of deduction.

The Diet

The 500-calorie diet is explained on the day of the second injection to those patients who will be preparing their own food, and it is most important that the person who will actually cook is present—the wife, the mother or the cook, as the case may be. Here in Italy, patients are given the following diet sheet.

Breakfast:	Tea or coffee in any quantity without sugar. Only one tablespoonful of milk allowed in 24 hours. Saccharin or Stevia may be used.
Lunch:	1. 100 grams of veal, beef, chicken breast, fresh white fish, lobster, crab, or shrimp. All visible fat must be carefully removed before cooking, and the meat must be weighed raw. It must be boiled or grilled without additional fat. Salmon, eel, tuna, herring, dried or pickled fish are not allowed. The chicken breast must be removed from the bird.
	2. One type of vegetable (only) is to be chosen from the following: spinach, chard, chicory, beet-greens, green salad, tomatoes, celery, fennel, onions, red radishes, cucumbers, asparagus, and cabbage.
	3. One breadstick (grissini) or one Melba toast.
	4. An apple, orange, a handful of strawberries, or one-half grapefruit.
Dinner:	The same four choices as lunch (above.)

The juice of one lemon daily is allowed for all purposes. Salt, pepper, vinegar, mustard powder, garlic, sweet basil, parsley, thyme, marjoram, etc., may be used for seasoning, but no oil, butter or dressing.

Tea, coffee, plain water, or mineral water are the only drinks allowed, but they may be taken in any quantity, and at all times.

In fact, the patient should drink about 2 liters of these fluids per day. Many patients are afraid to drink so much because they fear that this may make them retain more water. This is a wrong notion as the body is more inclined to store water when the intake falls below its normal requirements.

The fruit or the breadstick may be eaten between meals instead of with lunch or dinner, but not more than four items listed for lunch and dinner may be eaten at one meal.

To simplify this even further, you are allotted a MAXIMUM of:

- 2 protein servings
  - Raw: 100g / 3.5 oz
  - Cooked: 85g / 3.0 oz

- 2 fruit servings

- 2 vegetable servings
  - 4 cups total – no more than 2 cups in one sitting.

- 2 cracker servings

The list of foods allowed on the protocol is referenced above from Dr. Simeons' *Pounds &Inches.*

Participants with more fat find this amount of food is too much. We suggest protein servings are prioritized first, fruits and vegetables next, and lastly the cracker servings. Crackers should be prioritized if you wait too long, get too hungry, and need to restock blood glucose quickly.

Leaner participants find all of the food is needed, and hunger is manageable with precise timing for when and how much of the protocol food is eaten. For some who have less than 15 pounds to lose find they need a bit more food to keep hunger manageable.

Many misinformed participants make the mistake of assuming this part of the protocol is based on calories. Dr. Simeons used a pear and an apple as an example. A pear has fewer calories than an apple, yet the pear will create a stall in weight-loss. This might be due to the difference in how much the fruit stimulates leptin. The idea is to eat foods that supply adequate nutritional value, but reduces the hormones stimulus of leptin.

I do not suggest you try to make sense of why food is or is not on the protocol. It doesn't make sense with the old concept of calories and dieting, so unless you are willing to destabilize the controlled hormonal environment the protocol creates, don't eat more than the maximum Simeons recommended. In fact, you should eat only the amount necessary to remove hunger— prioritizing protein first.

Simeons listed two meals, which doesn't make sense with today's knowledge of how eating affects the hormonal system. We are going to recommend you eat only when hunger is adequate, and to stop when hunger is alleviated. To do this, we need to define the sense of hunger, and how a lack of leptin causes a sense of urgency to eat.

Here is the same described hunger/satiation scale I discussed in section 2. Again, it is a two-part scale used to measure and define the physical sense of urgency to eat, as well as the volume and symptoms of excess after hunger is gone.

**Remember, hunger is in terms of urgency, and
fullness is in terms of physical sensation.**

The hunger scale

1. *Disparaging.* Hunger is actually subsiding, as you feel less energy, less focus, and less desire for movement. Your headache continues, you feel lightheaded, and your stomach could have an acidic feeling. You're a bit cold, and your posture is lazy and rounded. You feel shaky, and a bit nauseous.

2. *Critical.* You have anger, irritability, your head hurts, and you don't care what food you eat as long as it's in large quantity—and fast (you're craving starches and sugars, combined with fat).

3. *Urgent.* You're uncomfortable, and you should have eaten ten minutes ago. Search for food is now imminent, as hunger is increasingly urgent, and choice of food is becoming less rational. Fast food is very appealing, and traditional restaurants seem less tolerable.

4. *Patient.* You're hungry, but can wait a bit. This is a good time to start prepping food for a meal. Most people can tolerate the wait in a restaurant at this point.

5. *Content.* You feel nothing, perfectly comfortable with or without food. Hunger is completely gone, with no sense of urgency.

The fullness scale

6. *Satisfaction.* You are confident your hunger is gone. You are feeling good.

7. *Satiation.* You're feeling a bit too satisfied, burping, and feeling some discomfort in the belly. Because you have no hunger, continuing to eat would mean that you've justified it emotionally. This is usually when a dieter feels guilty, and may defend a compensatory binge.

8. *Full.* You're uncomfortable, and definitely feeling your stomach. There's still some room for food because the stomach hasn't started

to stretch yet. Will need to wait three or more hours until your next meal.

9. *Discomfort*. You're very full, and feeling sick. Stomach is distended, with no more room for anything. Perhaps you have indigestion and a headache, and wish to lie down to reduce discomfort.

10. *Pain*. You've eaten so much you're contemplating inducing yourself to vomit in order to relieve the physical pain. You have to unbutton the first button on your pants, and can hardly stand to move. You're tired and need to take a nap, making it feel like being "Thanksgiving full."

This scale does not determine how fast hunger goes from feeling one number to the next. How quickly hunger elevates depends on the individual, and how much fat he or she has relative to activity level. The same goes for how long a person goes without feeling hunger, and how much food a person needs to feel satisfied.

Many who are obese don't feel hunger much at all, but when the physical urge to eat presents itself, hunger manifests rapidly and intensely. When food is eaten, it takes only small amounts of food, and satiation can last for longer periods of time.

For those who have less fat, it could take more time to adjust to hunger, but more food is needed to feel satisfied. This shows that with less fat, more food must be eaten to adequately stimulate enough leptin to desensitize the urge to eat sent from the brain.

Hunger is the communication to the conscious mind when food is hormonally necessary—and when it is not. During the very low-calorie protocol, it is imperative the participant eats according to this signal, and stops eating when the signal of hunger diminishes. The appropriate functional range for when to eat a snack would start at a 4, and end at a 5. When eating a meal, start at a 3.5 and end at a 5.5. Some people make the mistake of not eating past a 5, and find they feel slight hunger all day, and it increases in intensity by the end of the day.

It's important to make sure you finish your meals so that 20 minutes after you finish you notice you're at a 6—satisfaction.

Example VLCP Daily Log

What: *1cup coffee*
- Time: 7 am
- Hunger Before: 4.5
- Hunger After: 5

What: *½ apple, 1 cup tea*
- Time: 10 am
- Hunger Before: 4
- Hunger After: 5

What: *3 oz. water-packed tuna, 1 cup cucumber with sea salt, pepper, garlic, and vinegar, and the left-over ½ apple, 2 cups water*
- Time: 1 pm
- Hunger Before: 3.5
- Hunger After: 5.5

What: 1 cup salsa (tomato, onion, garlic, cilantro, sea salt, squeezed lime), 2 cups iced tea with lemon wedge
- Time: 4:30 pm
- Hunger Before: 3.75
- Hunger After: 5.0

What: 3.5 oz. cooked shrimp (weighed raw), steamed in 98% fat-free chicken broth, 2 cups lettuce, tomato, cucumber, 2 tbsp garlic vinegar dressing (recipe approved), 1 cup water
- Time: 6:30 pm

- Hunger Before: 3.5
- Hunger After: 5.5

What: ½ orange, 2 cups water
- Time: 9 pm
- Hunger Before: 4
- Hunger After: 5

You'll notice this individual needs to eat every 2-4 hours, avoided hunger below a 3.5, and didn't eat past a 5.5. You'll also notice they didn't need to eat their starch servings. Many people who are obese find all of what Simeons allowed is far too much. You should NOT force yourself to eat all of the food. Eating without hunger could cause a hormonal problem, and also takes away from the behavioral modification to eat, based on hunger.

By logging the time, and how much food is necessary to reach a 5-6 on the hunger scale, one can track how his or her hunger changes while he or she is losing the fat that produces leptin.

Leaner participants might notice hunger more often, and that they need all of the food Simeons suggested in order to relieve hunger. As fat loss brings a person's blood leptin level to where the protocol food is inadequate to stimulate enough leptin, hunger is felt more often, and persists after all of the protocol food has been eaten. At this point, if the individual has not yet met the three-week minimum, we suggest adding more protocol food to keep hunger within the 3.5–5.5 range so the entire twenty-day minimum can be reached.

Lotions and oils on the skin.

Dr. Simeons deducted that oil on the skin would have a negative effect on the outcome of the protocol. If things can absorb through the skin, and if fat indeed contains or can stimulate leptin, oil or fats on the skin could absolutely influence hormonal balance during the VLCP. This holds true especially for those who have more fat, and consequently, a higher susceptibility to over-stimulating leptin.

- Be careful to wash oils off hands if you're preparing food for other people. Use soaps that are not moisturizing.

- Make sure to use conditioners only on the ends of your hair, and do not get your hair dyed during the VLCP.

- Avoid oils used during massage, and on the hands and feet during manicures and pedicures.

- Your powder foundation and liquids for makeup need to be oil-free. Mascara and lipstick should not cause a problem.

- You can use aloe Vera gel, products with mineral oil such as baby oil, and other oil-free products.

Artificial sweeteners and supplements.

Artificial sweeteners: Artificial sweeteners have been shown to stimulate leptin, even though they have no calories. Participants who have more fat should be cautious using these sweeteners as you have more leptin that can be over-stimulated and can cause fat gain.

If you choose to add diet soda or other artificially sweetened drinks, make sure you're at least hungry to offset any potential leptin problems. Leaner participants seem not to be as influenced by these weak leptin stimulators. The same goes for too much caffeine. Be cautious!

Supplements: Supplements that include fish oils or omega fatty acids should not be used during the VLCP. Other supplements that contain water-soluble vitamins can be continued.

Medications, blood pressure, gout, and Type II diabetes.

Medications: You need to disclose all prescribed medications prior to starting the hCG protocol. All medications should continue as directed by your prescribing doctor. Discuss the hCG protocol with your prescribing doctor to question symptoms you are medicating, and whether fat loss/weight would influence your dosage. As you lose significant amounts of fat, you will need to revisit your doctor to reassess what you are taking, and if your medication is still appropriate.

Blood Pressure: Most people experience low blood pressure, especially at the end of the VLCP. An abnormal dip in blood pressure (below 110/70) usually occurs at the end of one's protocol, and could indicate the need to transition into the second phase. Those who are on prescribed medication for moderately high blood pressure might experience a drop in blood pressure that should be discussed with their prescribing doctor. Check your blood pressure daily.

Gout: If you have bouts with gout, you will definitely experience symptoms during the VLCP. Make sure you discuss this with your doctor, as a prescription for these symptoms will help them immediately.

Type II diabetes: Fat and high levels of leptin are highly linked to Type II diabetes. Before starting the hCG protocol, it is important you discuss this form of fat loss with your providing doctor, and see that you are monitored closely. As significant amounts of fat are lost (and subsequent blood leptin levels fall), you will need to be reevaluated. Make consistent appointments throughout your protocol with your doctor.

Exercise, pregnancy, and constipation

Exercise: During the VLCP, weight-bearing exercise or intense exercise that requires an increase in protein, should be avoided. Otherwise, you will experience protein deficiency, and will feel lethargic.

Moderate cardio, stretching, yoga, and other forms of postural exercise are more appropriate. Too much exercise can have the reverse affect during the protocol, making the hCG ineffective to stimulate enough leptin to protect against starvation.

Pregnancy: The chance of becoming pregnant is higher, so make sure you use protection or abstain from sex during the protocol.

Constipation: Constipation is a very common side effect of eating less. It's normal to have small bowel movements every 2-3 days, but longer than 4 days could mean you have a problem.

It's not because of lack of fiber, but rather there isn't much food to stimulate natural peristalsis (smooth muscle contractions) that pushes the waste out of the body. If you experience constipation normally, you will definitely need to be proactive during the VLCP.

Drinking yerba matte tea or a stimulant type of tea daily should help. If not, a more powerful stimulant can be used (Milk of Magnesia, etc.). Consider asking about a suppository at your local pharmacy. Remember: be proactive, and don't wait until it becomes a big problem.

Menstruation

For females, the best time to start the protocol is just after finishing her period. If that isn't possible, start any time. She could have an increase in hunger prior to her period, and a dramatic decline in hunger during ovulation. These hunger fluctuations will continue after the protocol.

Watch for increased symptoms of PMS during the VLCP. Even some men experience similar symptoms (except cramping).

Women that have partial hysterectomies might notice these hunger fluctuation, as well as women who have IUDs. Women who are in menopause could even have a period or spotting during the VLCP, or notice that hot flashes completely go away during this time.

Dr. Simeons mentioned that hCG is not necessary during menstruation. Some women continue with hCG, and others don't. This is a personal decision that needs trial and error for the individual to decide. Some women reduce their hCG dose during this time instead.

Deviations or "cheating" on the VLCP

Cheating on the VLCP is when 1) protocol-approved food is eaten without hunger or eaten until full, 2) protocol-approved food is eaten in excess of what's prescribed, 3) food or drink that is not listed is consumed, or 4) excessive exercise is continued.

Cheating on the VLCP or the second phase will cause an overreaction and inflammation from the thyroid, pancreas, gall bladder, and endocrine system. Diarrhea, stomach pain, fluid retention, hot flashes, sleep interruption, irritability, and definite fat gain could be the result.

Once the protocol is followed for enough days to make up for the imposed damage, the body will reestablish a consistent and predictable homeostasis. If a participant is consistently cheating, the first phase should transition into the second phase, and the emotional dependency on food should be evaluated. Maybe the participant needs to wait a period of time before reassessing their desire for emotional autonomy from food before attempting the protocol again.

Due to the effects that hCG and the protocol have on the entire endocrine system, emotional dependency with food and eating should be a high priority, and assessments should be implemented during the pre-protocol evaluations. An emotional dependence on eating or a food addiction should be considered prior to receiving a prescription. We highly encourage the participant to receive appropriate therapy and guidance from a therapist or counselor.

Breaks in the VLCP

Many people have events, traveling interruptions, or intentional breaks during the VLCP. This means that the participant transitions into the second phase, follows the second-phase boundaries, and reverts back into the VLCP with no more than a two-week break. The longer the break, the more one will need to "load" to start the VLCP again. If the break is less than a week, the participant should start the VLCP the same day he starts using the hCG again. Breaks longer than a week need a "mini-load" of one, or possibly two days of forced-feeding.

If a break is past the fifth week of the VLCP, the entire second phase and maintenance period should be fully executed before a new round of the hCG protocol is started.

THE SECOND PHASE

CAUTION: Due to the lack of fat and the severely reduced quantity of food during the VLCP, you have a greater risk of experiencing gall bladder inflammation, pancreatitis, dumping syndrome, and other painful side effects. It is very important that fat is introduced slowly and with caution. DO NOT start the first day of the second phase with bacon, sausage, hamburger, or other lard-laden foods. Eat food slowly, chew sufficiently, and be very careful to keep portions appropriate for the hunger scale. The goal is to continually maintain hunger between 3.5-5.5—and with foods that will be tolerated by the stomach and digestive system.

The second phase of the protocol begins the fourth day after the last injection, and lasts three weeks. This phase is as fundamentally important as the first because it allows the recovery and hormonal stability necessary for the endocrine system to readjust to the new level of fat and resulting fat hormones.

During the first phase, as excess fat declines at a steady linear rate, the result is a steady decline in blood leptin. Even though large fluctuations are controlled during the first phase, this overall decline in blood leptin, and the extraordinary demand on the fat cell's mitochondria, require rest and recovery in order for a metabolic and hormonal adaptation to occur. This alteration within the brain and organs of the endocrine system requires hormonal consistency over a period of time. This is accomplished by continuing to monitor and control hunger between 3.5 and 5.5, and by avoiding foods that would strongly stimulate leptin, such as sugar and starch.

If you have a large fluctuation in leptin at this stage, the endocrine system cannot adapt or readjust to the new amount of body fat. This is like attempting to hit a bull's eye on a moving target. The other organs need predictability and consistency for an appropriate hormonal adjustment. Because this system was previously balanced with the large amount of leptin, it could react to instability by over- or under-stimulating other organs causing the creation of more leptin. This would cause a very fast regain of the fat that was just lost.

The hormonal adaptation not only requires constant attention and monitoring of hunger, but also attention to what types of foods are eaten. It's vital to stay away from foods that are sensitive to fast absorption, or that are high in sugar or starch content. Because leptin is stimulated as blood glucose levels rise, it's extremely important you keep your leptin and blood glucose levels from falling too low or rising too high. This does not mean you can't have carbohydrates, but you are limited to choices that are low on the glycemic index, or that have minimal sugar.

Here is what Dr. Simeons wrote:

Concluding a Course

When the three days of dieting after the last injection are over, the patients are told that they may now eat anything they please, except sugar and starch, provided they faithfully observe one simple rule. This rule is that they must have their own portable bathroom scale always at hand, particularly while traveling. They must without fail weigh themselves every morning as they get out of bed, having first emptied their bladder. If they are in the habit of having breakfast in bed, they must weigh before breakfast.

*It takes about 3 weeks before the weight reached at the end of the treatment becomes stable, i.e., does not show violent fluctuations after an occasional excess. During this period, patients must realize that the so-called carbohydrates that are in sugar, rice, bread, potatoes, pastries, etc., are by far the most dangerous. If no carbohydrates whatsoever are eaten, fats can be indulged in somewhat more liberally, and even small quantities of alcohol, such as a glass of wine with meals, does no harm, but **as soon as fats and starch are combined, things are very liable to get out of hand.** This has to be observed very carefully during the first three weeks after the treatment is ended, otherwise disappointments are almost sure to occur.*

Again, I do not suggest you carry a scale with you, nor should you obsess over your weight. Weighing once a week is fine, but a better guide would be observing how your clothes fit. If you focus on something, it should be your ability to recognize and assess hunger. Before you eat you should first, make sure you are physically hungry, second, make sure the food isn't a potent hormonal stimulant like sugar or starch, and third, avoid any sense of fullness.

To keep it simple, stay away from bread, crackers, soda, sugar, candy, cake, popcorn, bananas, pineapple, cereal, rice, potatoes, sweetened yogurt, etc. Look up "glycemic index" to find a list of common foods that are less starchy than others. During the second phase you may add more types of meat, dairy, nuts, eggs, and other fruits and vegetables. Avocados, peanuts, salmon, capers, artichokes, berries, feta, cheddar, unsweetened Greek yogurt, and oil dressings are now acceptable food choices—as long as they are eaten slowly, and only to remove hunger.

This phase is extremely important for the body to acclimate and regulate energy balance without hCG, and the controlled environment of the very low-calorie protocol, to your new level of fat hormones. Dr.

Simeons believed this phase re-regulates the brain and body to a new and improved metabolic rate.

This metabolic increase will not be felt until after the three weeks are over. During the second phase, the body gets some vital rest needed after the extreme labor involved when supplying the body with fuel, without food. For a person who lost 20 pounds of fat during a six-week first phase, that expenditure is similar to running 700 miles in the same amount of time! Maybe his or her muscles are not fatigued, but the fat cells are.

This incredible demand on fat metabolism could up-regulate the mito-chondrial gene, creating new, smaller mitochondria surrounding the leftover fat cells. Mitochondria are the engines of fat cells, churning and creating fuel out of fat. This metabolic regulatory system is primarily controlled by leptin and energy demand.

During the first phase, both your leptin and energy demand increase, greatly challenging the mitochondria so that it stimulates the develop-ment of more mitochondria. This development and growth could occur during the second phase, as the body is resting from deprivation. The new mitochondria add to the body's burning capacity, allowing for a higher potential fat-burning capacity, and improving overall resting energy expenditure.

To make matters even more amazing, the loss of significant fat elimi-nates equal stimulus of leptin—removing symptoms associated with excessive leptin levels, and decreasing the risk of disease. As the body recovers during the second phase, the endocrine system acclimates to the new level of fat hormones—affecting hunger signal, thyroid func-tion, menstruation, sex drive, sleep, and more. This acclimation can be disrupted with large surges of leptin and blood glucose, and are poten-tially the reason Simeons observed the sugar and starch ban during the three-week period, after the first phase.

Caution: eat slowly.

Glucose is a direct stimulant of leptin, and creates a fast and powerful impulse of fuel, not only from the immediate absorption of the sugar, but from leptin's stimulated metabolism of fat energy. This is a double whammy. It's imperative any food consumption during the second phase is done slowly, cautiously, and without sugar or starch.

The concept is to keep blood glucose confined to a small, controlled range. This is manipulated by strategic timing of when to eat, slowly eating what stimulates leptin, and knowing when to stop. This requires awareness of what's in the food you're eating, and where you are on the hunger scale at all times, which determines when and how much to eat. Remember, when leptin is elevated, hunger subsides.

Essentially, if you're not hungry, your fat metabolism is elevated, and you have no physical need for food.

The second-phase boundaries are vital in keeping blood leptin fluctuations slow and controlled. This ability is dependent on your sharp awareness of the body's signals of hunger. The second phase is a test of your keen sense and ability to direct food consumption, coordinated with the appropriate signal of hunger, as urgency to eat rises and falls. If you have a large amount of fat yet to lose, this signal is still very sensitive and will be—until the extra fat is gone.

People who are largely obese often observe they are not hungry for long periods of time; then hunger presents itself very quickly. This abrupt change, with very little warning, demands urgent attention. If this describes you, I suggest you pack a cooler of food so something is available at the first sign that your hunger is increasing. Eat slowly in case you have an abrupt change in fullness as well. It could take very little to remove the hunger, so eat cautiously slow.

During the second phase you are attempting to replicate the same hormonal balance you had during the first phase. You need to eat foods that are slow to absorb, and eat them slowly, in order to effectively evaluate whether or not your hunger is losing strength. When you eat too fast, it's difficult to clearly feel when you're satisfied, and how much you've eaten. For many who eat quickly, by the time they are done, they've had more than their body needs, and their satiation doesn't show up until it's too late.

Cheating on the second phase

Cheating on the second phase is done by eating when you have no hunger, overeating until full, and eating foods that are potent stimulators of leptin (sugar and starches). Many who cheat and test the second-phase boundaries gain fat faster than makes caloric sense. The oversensitivity to stimulate leptin might be due to the past regulation of the endocrine system with more fat (previously there before the VLCP). This means that your body will gain fat back until you have an amount similar to what you had before starting the protocol.

It could take the body three weeks of hormonal consistency to acclimate to the new level of fat hormones. Keeping leptin levels consistent with the new amount of fat would be a fundamental necessity for the hormonal acclimation, and for "re-setting" the endocrine system. Cheating negates this acclimation, and promotes reversal to the old system's equilibrium, which requires fat to accumulate in order to provide the required excess leptin.

Calories consumed: relative to fat and expenditure, indicated through hunger.

Now that we know fat metabolism determines how much food is needed, caloric needs are relative to how much fat your body uses for fuel. If you have too much leptin, your body is using more fat fuel, and doesn't hormonally need food.

Basically, the higher percentage of body fat you have, the less food your body needs hormonally. As you lose the fat, and your body returns to its normal, ideal state, it will produce less leptin, and your body will hormonally need more food. Based on this theory, if you eat a predetermined amount of food, and override physical hunger or satiety, you're setting yourself up for regaining the fat you just lost.

We recommend keeping track of your hunger scale during the day, recording the time, the food, and how much your body needs to remove hunger. Then calculate the calories after the fact. This will give you an idea of how your body rehabilitates after each protocol, not only removing unwanted fat hormones, but improving your fat-burning capacity (with additional mitochondria). This is unique to each person's body. No single diet can determine an individual's relative needs for food. This is why the hunger scale is essential in helping you eat to match your body's needs.

Refer to the hunger/satiation scale described earlier in this chapter. This two-part scale is used to measure and define the physical sense of urgency to eat, as well as the volume and symptoms of excess after your hunger is gone. Ideally you confidently can assess this hunger scale before and after you eat. Throughout the process, participants should use their own words and descriptions.

**Remember, *hunger* is in terms of urgency, and
fullness is in terms of physical sensation.**

During the first phase, you hopefully created your own defined hunger/ satiation scales, using words and descriptions for your physical cues. If not, you can continue to use what we've put together, redefining those numbers as you feel necessary. The most important assessments are where you are on these scales before you eat, and being able to determine when to stop eating. Once the second phase is done, you can slowly return sugars and starches to what you eat, continuing the same observation as to how they influence speed of satiation, as well as how much time it takes to feel an urge to eat again.

Preparation is extremely important during the second phase of the protocol. And when going back into the real world, it's especially vital for those with busy schedules. Making sure you have fruit, snacks, and prepared food when hunger presents itself, is key in controlling your eating schedule. As you understand your own body, you may prepare food that you know has longer lasting effects, or short effects, based on what you need.

Preparation is a part of making your physical health and taking care of your body a priority, just as you would a child.

This approach to eating completely personalizes when and how much to eat, all the while keeping track of what food affects your appetite throughout the day. For example, we suggest you eat breakfast when you feel hungry. For many of us, this isn't until a couple of hours after waking, thus you may make lunch a smaller meal. Some people who exercise notice their hunger changes after their workout, with either an increasing or decreasing appetite. The choice to eat is completely based on hunger, and ending a meal is based on *removal of hunger, not fullness.*

Eating based on the hunger scale reveals your emotional drive, and reinforces functional eating. The more aware you are, the more obvious emotional justifications become—whether it's eating when you have no hunger, or continuing to eat after hunger is gone. When hunger is removed, that is the sign leptin is elevated, and blood glucose is back to its perfect range. What I love about this concept is that it's completely individualized and tailored to your body. This eliminates judgment towards food and rules that are difficult to apply.

The ability to identify the inherent sense of hunger and satiety, and the ability to discern psychological hunger from physical hunger, are the foundation for living free of dieting, deprivation, and gluttony. When the hunger scale is combined with the knowledge gained from understanding hormone regulation, a more dynamic, adaptable, and personalized diet can be followed. Once you fully understand how your body responds to food (based on hunger), confidence is rebuilt, and fear of weight gain dissolves.

Confidence is built as the body becomes predictable. This predictability is learned by tracking the body's hunger signals using time, what, and how much you eat—and limiting yourself between 3.5 and 5.5.

Here is the same example log at the end of the day using the hunger scale.

What: Coffee with cream
- Time: 7:00am
- Hunger Before/ Hunger After: 4.5/5.0
- Calories:50

What: Lunch
- Time: 1:30pm
- Hunger Before/ Hunger After: 3.5/5.5
- Calories:445

What: Breakfast
- Time: 10:00am
- Hunger Before/ Hunger After: 3.5/5.0
- Calories:250

What: Snack
- Time: 3:45pm
- Hunger Before/ Hunger After: 4.0/5.0
- Calories:200

What: Snack
- Time: 11:35am
- Hunger Before/ Hunger After: 4.0/4.75
- Calories:85

What: Dinner
- Time: 6:00pm
- Hunger Before/ Hunger After: 3.5/5.5
- Calories:385

After following hunger throughout the day, this person only needed to eat 1415 calories.

How much fat a person has greatly influences when they experience hungry, how much food is necessary to remove that hunger, and how long it will take before they will need to eat again.

Many people using the hunger scale end up eating less than anticipated, and feel more energized throughout the day. The goal is to stay within a small range on the hunger scale, avoiding both low numbers and high numbers alike. Your goal is to do the same thing as a child who is too busy to eat until his hunger becomes uncomfortable—he stops playing, eats just enough for the hunger to go away, and continues what he was doing before.

To make eating functional, allow yourself to eat what you want without judgment, and limit quantity, using the body's signal.

Acknowledgments

First I'd like to acknowledge and thank my development editor, Connie Anderson of Words and Deeds, Inc.; technical editor, Julianne Kirry; and keen-eyed proofreader, Diane Keyes. Without their talents, this book would have been only twenty pages long, extremely repetitive, hard-to-understand, offensive, and boring. They took what was in my head, and helped me paint a picture that you could understand and apply. I am so grateful I found the right people.

I'd like to acknowledge and thank Drs. Michael Halliday, Diane McConnehey, and Heidi Anderson. They trusted my work, and allowed me to monitor hundreds of their patients.

I am deeply grateful to Boise State's Medical Epidemiologist, Dr. Lee Hannah for evaluating the resting metabolic rate results. She gave relevance to the most telling data I've collected.

Boise State University required I take over two years of physiology as well as two years of chemistry. Without such rigorous educational standards, I couldn't have understood the research that provided the basis for my hypothesis, specifically to the Kinesiology department. Thank you for having such passion for the health of the human mind and body, and for also allowing me to use the human performance lab for testing.

A special thank you to Denise Watson and Dr. Ed Hagen from Vivify HCG Weight Loss in Edina, Minnesota. You've given me the opportunity to teach your patients, one by one, for over a year. Repeating the same explanation over and over, hundreds of times, has been an invaluable part of refining how I teach the protocol process to participants. This work with you has been precious.

I also want to express my gratitude for Dr. Mayer Eisenstein, who took the time to review and write the foreword. I never thought I'd come across another person who has fallen in love with the human body as deeply as I have. Thank you for your energy, passion, and integrity.

I'd like to acknowledge my team at Tone. You were there for me when I started, grew, expanded, succeeded, failed, lost, and when I gave up my creation and training career for the hCG protocol. Each of you has played a profound part in my life, and I am grateful. I truly love you all and everyone else who voluntarily submitted themselves to my craziness, and were loyal to me. I love you all.

My hard-working and humble parents—for being the ultimate examples of perseverance and integrity. To sister Becky for introducing me to the hCG protocol, listening to my ignorant and skeptical rant, and then forgiving me. To sister Katie for your incredible artistic mind, and for the book title and cover design. To them and all my other siblings—Steve, Laura, Daniel, Jennalee, Debbie, Melissa, Cliff, Mike, Big Jeff, Little Jeff, and Jean —for helping develop my "character."

References

(1) Lijesen, G.K.S., et al. 1995. The effect of human chorionic gonadotropin (HCG) in the treatment of obesity by means of the Simeons' therapy: a criteria-based meta-analysis. *The British Journal of Clinical Pharmacology.* 40:237-243

(2) Zhang, Y., et al. 1994. Positional cloning of the mouse *obese* gene and its human homologue. *Nature.* 372: 425-432

(3) Morton, G.J. 2007. Hypothalamic leptin regulation of energy homeostasis and glucose metabolism. *J Physiology.* 583.2: 437-443

(4) N.Y. Ann. 2002. Regulation of fat metabolism in skeletal muscle. *National Academy of Sciences of the USA.* 967:217-235

(5) Margetic, S., et al. 2002. Leptin: a review of its peripheral actions and interactions. *International Journal of Obesity.* 26: 1407-1433

(6) Groschl, M., et al. 2001. Identification of Leptin in Human Saliva. *The Journal of Clinical Endocrinology & Metabolism.* 86(11): 5234-5239

(7) Mars, M., et al. 2006. Fasting leptin and appetite responses induced by a 4-day 65% energy-restricted diet. *International Journal of Obesity.* 30: 122-128

(8) Rosenbaum, M., et al. 2002. Low dose leptin administration reverses effects of sustained weight-reduction on energy expenditure and circulating concentrations of thyroid hormones. *The Journal of Clinical Endocrinology & Metabolism.* 87:2391-2394

(9) Flier, J.S., et al. 2000. Leptin, nutrition, and the thyroid: the why, the wherefore, and the wiring. *The Journal of Clinical Investigation.* 105(7): 859-861

(10) Ahima, R.S., et al. 1996. Role of leptin in the neuroendocrine response to fasting. *Nature.* 382: 250-252

(11) Kahn, M.Y. 2003. Role of AMP-activated protein kinase in leptin-induced fatty acid oxidation in muscle. *Biochemical Social Transactions.* 31(pt 1): 196-201

(12) Flier, J.S. 1998. Clinical Review 94: What's in a name? In search of leptin's physiologic role. *Journal of Clinical Endocrinology and Metabolism.* 83(5):1407-1413

(13) Wolfgang, M.J., et al. 2007. Regulation of hypothalamic malonyl-CoA by central glucose and leptin. *The National Academy of Sciences of the USA.* 104(49): 19285-19290

(14) Qian, H., et al. 1998. Leptin regulation of peroxisome proliferator-activated receptor-gamma, tumornectrosis factor, and uncoupling protein-2 expression in adipose tissue. *Biochemical and Biophysical Research Communications.* 246(3): 660-667

(15) Wolfgang, M.J., et al. 2008. Hypothalamic malonyl-CoA and the control of energy balance. *Molecular Endocrinology.* 22(9): 2012-2020

(16) Orci, L., et al. 2003. Rapid transformation of white adipocytes into fat-oxidizing machines. *The National Academy of Sciences of the USA.* 101(7): 2058-2063

(17) Islama, D., et al. 2003. Modulation of placental vascular endothelial growth factor by leptin and hCG. *Molecular Human Reproduction.* 9(7): 395-398

(18) Maymo', J.L., et al. 2009. Up-regulation of placental leptin by human chorionic gonadotropin. *The Endocrine Society.* 150(1): 304-313

(19) Linnemann, K., et al. 2000. Leptin production and release in the dually *in vitro* perfused human placenta. *The Journal of Clinical Endocrinology & Metabolism*. 85(11): 4298-4302

(20) Islami, D., et al. 2003. Possible interactions between leptin, gonadotrophin-releasing hormone and human chorionic gonadotrophin. *European Journal of Obstetrics, Gynecology, and Reproductive Biology*. 110(2): 169-175

(21) Considine, R.V., et al. 1996. Serum immunoreactive-leptin concentrations in normal-weight and obese humans. *The New England Journal of Medicine*. 334(5): 292-295

(22) Friedman J.M., et al. 1998. Leptin and the regulation of body weight in mammals. *Nature*. 395:763-770

(23) Cha, S.H., et al. 2005. Inhibition of hypothalamic fatty acid synthase triggers rapid activation of fatty acid oxidation in skeletal muscle. *The National Academy of Sciences of the USA*. 102: 14557-14562

(24) Schwarts M.W., et al. 2006. Central nervous system control of food intake. *Nature*. 404: 661-671

(25) Cha, S.H., et al. 2003. Hypothalamic malonyl-CoA as a mediator of feeding behavior. *The National Academy of Sciences of the USA*. 100: 12624-12629

(26) Cha, S.H., et al. 2006. Hypothalamic malonyl-CoA triggers mitochondrial biogenesis and oxidative gene expression in skeletal muscle: role of PGC-1α. *The National Academy of Sciences of the USA*. 103: 15410-15415

(27) Thupari, J.N., et al. 2002. C75 increases peripheral energy utilization and fatty acid oxidation in diet-induced obesity. *The National Academy of Sciences of the USA*. 99: 9498-9502

(28) Lederman, S.A., et al. 1999. Maternal body fat, water during pregnancy: do they raise infant birth weight? *American Journal of Obstetrics and Gynecology*. 180: 235-240

Glossary

Hypothesis: A tentative statement that proposes a possible explanation to some phenomenon or event. A useful hypothesis is a **testable** statement that may include a prediction.

Theory: A general explanation based on a large amount of data.

Protocol: A medical guideline.

Homeostasis: The ability to maintain internal equilibrium by adjusting a physiological process, Homeostasis is a result when a condition of properties are stable and constant.

Metabolism: The sum of the physical and chemical processes in an organism, which is how material substance is produced, maintained, and destroyed, and how fuel is made available.

Substrates: The material or substance on which an enzyme acts.

Glucose: A carbohydrate that is initially synthesized by chlorophyll in plants using carbon dioxide from the air and sunlight as an energy source. In humans glucose is a monosaccharide that circulates in the blood at a concentration of 65-110 mg/mL of blood.

Glycogen: The storage form of glucose in animals and humans, which is analogous to the starch in plants. Glycogen is synthesized and stored mainly in the liver and the muscles. It is an anaerobic fuel source used when there is powerful fuel demand, such as sprinting.

Organelle: An organelle is one of several structures with specialized functions, suspended in the cytoplasm of a cell.

Mitrochondria: An organelle that provides the fuel that cells need to function. They are the "gas pumps" that create and provide the adenosine triphosphate (ATP) to fuel the many functions of the body.

Adenosine Tri Phosphate (ATP): The high-energy molecule that that fuels just about everything we do. It is present in the cytoplasm and nucleoplasm of every cell.

Hypothalamus: A region about the size of a pearl, located in the middle of the base of the brain. It controls an immense number of bodily functions. The hypothalamus plays a vital role in maintaining balance within the endocrine and nervous systems.

Diencephalic: The posterior part of the forebrain that connects the midbrain with the cerebral hemispheres, encloses the third ventricle, and contains the thalamus and hypothalamus.

Endocrine System: A system, or network of organs, that is made up of glands that produce and secrete hormones that regulate the body's growth, metabolism (the physical and chemical processes of the body), and sexual development and function. The hormones are released into the bloodstream and may affect one or several organs throughout the body. Like a thermostat that regulates temperature in a room, the endocrine system is regulated by a similar feedback within each organ.

Hormones: Chemical messengers created by the body that transfer information from one set of cells to another. They coordinate the functions of different parts of the body.

Leptin: A protein hormone that plays a key role in regulating energy intake and energy expenditure, including appetite and metabolism. It is one of the most important adipose derived hormones.

Gonadotropins: Protein hormones secreted by the pituitary gland. These hormones (follitropin (FSH), lutropin (LH), and placental chorionic gonadotropins (hCG)) are central to the complex endocrine system that regulates normal growth, sexual development, and reproductive function. The hormones LH and FSH are secreted by the anterior pituitary gland, while hCG is secreted by the placenta.

Adrenals: Endocrine glands that sit atop the kidneys. They release hormones in response to stress through the synthesis of corticosteroids such as cortisol and catecholamines such as epinephrine. They cause the physical "fight or flight" response from the body stressful situations.

Thyroid: One of the largest endocrine glands that is found in the neck. It controls how quickly the body uses energy, makes proteins, and controls how sensitive the body should be to other hormones. The principal hormones the thyroid produces are triiodothyronine (T_3) and thyroxine (T_4). These hormones regulate and affect the rate of metabolism and the rate of function of many other systems in the body.

Anorexigenic: Describes the character of a hormone or substance that causes loss in physical hunger.

Body fat composition: A percentage that estimates a measure of how much a person's total weight is fat. It is measured with skinfold or caliper tests, bioelectrical impedance analysis (BIA), air displacement with a Bodpod, and/or water displacement with hydrostatic weighing. These tests are based on the two-component (fat and fat-free mass) model of body composition.

Glycemic index: A measure of the effects of carbohydrates on blood sugar levels.

Polycystic ovarian syndrome: A condition in which there is an imbalance of a woman's female sex hormones that may cause skin changes, changes in the menstrual cycle, small cysts in the ovaries, trouble getting pregnant, and other problems. This condition is thought to be caused by high blood levels of both insulin and leptin, which polarize the gonadotropin (LH/FSH) ratio from the ovaries. The consequence is abnormal increases in testosterone production in females.

Gout: A form of arthritis that occurs when uric acid builds up in blood and causes joint inflammation.

Fatty Acid Synthase (FAS) Inhibitor: A pharmacological or genetic tool that modifies the activity of different enzymes regulating fatty acid metabolism and supports the regulation of feeding.

Pancreatitis: When the pancreas gets inflamed when pancreatic enzymes that digest food are activated in the pancreas instead of the small intestine.

Dumping syndrome: A condition where ingested foods bypass the stomach too rapidly enter the small intestine largely undigested.